AF352981

# Modern Research of Gears and Power Transmission

# Modern Research of Gears and Power Transmission

Editors

**Aleksandar Miltenovic**
**Franco Concli**
**Stefan Schumann**

Basel • Beijing • Wuhan • Barcelona • Belgrade • Novi Sad • Cluj • Manchester

*Editors*

Aleksandar Miltenovic
University of Niš
Niš
Serbia

Franco Concli
Free University of
Bolzano/Bozen
Bolzano
Italy

Stefan Schumann
TUD Dresden University
of Technology
Dresden
Germany

*Editorial Office*
MDPI
St. Alban-Anlage 66
4052 Basel, Switzerland

This is a reprint of articles from the Special Issue published online in the open access journal *Applied Sciences* (ISSN 2076-3417) (available at: https://www.mdpi.com/journal/applsci/special_issues/E449E1E9E7).

For citation purposes, cite each article independently as indicated on the article page online and as indicated below:

Lastname, A.A.; Lastname, B.B. Article Title. *Journal Name* **Year**, *Volume Number*, Page Range.

**ISBN 978-3-7258-1279-0 (Hbk)**
**ISBN 978-3-7258-1280-6 (PDF)**
**doi.org/10.3390/books978-3-7258-1280-6**

Cover image courtesy of Aleksandar Miltenovic

# Contents

*Article*

# Investigation of the Influence of Contact Patterns of Worm-Gear Sets on Friction Heat Generation during Meshing

**Aleksandar Miltenović *, Milan Banić, Nikola Vitković, Miloš Simonović, Marko Perić and Damjan Rangelov**

Faculty of Mechanical Engineering, University of Niš, 18000 Niš, Serbia; milan.banic@masfak.ni.ac.rs (M.B.);
nikola.vitkovic@masfak.ni.ac.rs (N.V.); milos.simonovic@masfak.ni.ac.rs (M.S.);
marko.peric@masfak.ni.ac.rs (M.P.); damjan.rangelov@masfak.ni.ac.rs (D.R.)
* Correspondence: aleksandar.miltenovic@masfak.ni.ac.rs

**Abstract:** Friction losses and scuffing failures are interesting research topics for worm gears. One of the factors leading to scuffing is the heat generated in the contact of gear teeth. The contact geometry of worm gears is complex, leading to high friction between contact surfaces. High friction between contact surfaces during operation generates heat friction that causes the occurrence of scuffing, which in turn determines the scuffing load capacity. To analyse the thermal characteristics of a worm-gear pair and the thermal behaviour of contact teeth, a direct-coupled thermal–structural 3D finite element model was applied. The heat flux due to friction-generated heat was determined on the gear tooth to investigate thermal characteristics and predict transient temperature fields. This study permits an in-depth understanding of the temperature fields and the friction heat generation process. Also, better control of the contact pattern between worm-gear teeth would decrease friction heat and increase scuffing load capacity. This paper investigates the transient thermal behaviour among different pinion machine setting parameters that can result in an optimal tooth-contact pattern that produces a lower temperature field, thus achieving higher transmission efficiency.

**Keywords:** worm gear; friction heat generation; FEM; contact pattern

**Citation:** Miltenović, A.; Banić, M.; Vitković, N.; Simonović, M.; Perić, M.; Rangelov, D. Investigation of the Influence of Contact Patterns of Worm-Gear Sets on Friction Heat Generation during Meshing. *Appl. Sci.* **2024**, *14*, 738. https://doi.org/10.3390/app14020738

Academic Editor: Adrian Irimescu

Received: 24 November 2023
Revised: 10 January 2024
Accepted: 13 January 2024
Published: 15 January 2024



## 1. Introduction

Worm gears find extensive use in various industries and have a significant share in the gear-and-drives market. They enable the attainment of significant gear ratios, reaching up to 300:1 in a single gear stage. These gears are capable of transmitting motion at a 90° axis angle, have a compact design, operate quietly, and are commonly self-locking, preventing reverse driving [1].

The load-capacity calculation of worm gears is standardised in ISO/TS 14521 [2] and DIN 3996 [3]. These standards cover calculations of load capacity for worm gears, such as wear, pitting, worm-shaft bending, tooth breakage, and temperature.

Höhn [4] developed a method that allows the determination and optimisation of a worm and worm-gear contact pattern. This method is based on a point-by-point simulation of a worm and worm gear that considers the manufacturing process. Simon [5] developed a method for computer-aided loaded-tooth contact analysis. This method can analyse different types of cylindrical worm gears, and it covers contact of the theoretical line and contact in point. Sharif [6] presented a wear model for the prediction of wear patterns. This model is based on an electrohydrodynamic lubrication that provides wear distribution on the tooth surfaces during meshing.

Paschold [7] presented a method for determining the efficiency and heat balance of gearboxes with worm gears. For this purpose, he developed new approaches for the calculation of losses under load and without load. Also, he developed new algorithms for normalisation and node linking and customised formulas for thermal resistance. Roth [8] presented a numerical temperature simulation for worm gears that considers transient multidimensional heat transfer and local frictional loading due to contact. He gave a simplified calculation of worm-gear contact temperature. Stahl [9] presented a new model for determining pitting load resistance that considers contact patterns. Hitcher [10] carried out a numerical simulation to study the influence of cutting parameters and to find the optimal manufacturing procedure. Daubach [11] developed a simulation model for the abrasive wear of worm gears based on the energetic wear equation. This simulation model includes tooth contact analysis and tribological calculation to determine friction and wear. Tošić investigated the thermal effects of slender EHL contacts in the case of disc–disc contact [12] and contact of worm gears [13]. He introduced a numerical procedure for the investigation of the transient thermal EHL contact of worm gears with nonconjugated meshing action. He used FEM by implementing the thermal elastohydrodynamic lubrication model.

Chernets [14] evaluated a computation method for worm gears with Archimedean and involute worms in operation. His approach is based on contact pressures, wear at worm-gear teeth, gear life, and sliding speed. Jbily [15] proposed a numerical model for predicting the wear of worm gears. This numerical model is based on Archard's wear formulation and considers the influence of lubrication on the local wear coefficient, which depends on the minimum lubricant film and the amplitude of surface roughness. Oehler [16] developed a new calculation method for the prediction of the efficiency of worm-gear drives that is compatible with DIN 3996 and that could be incorporated into standard calculations. His research is based on simulations and validated by experiments, and his approach allows engineers to compare different drivetrain concepts regarding efficiency. Miltenovic [17] investigated the thermal design and prediction of the whole worm-gear drive using FEM.

Li [18] proposed a scuffing model for spur-gear contacts. For this purpose, a heat transfer formulation was devised to evaluate gear bulk temperature. Bulk temperature was used for EHD modelling to determine tribological behaviour within the contact zone. Yanzhong [19] investigated the friction heat generation during meshing of spiral bevel gears for different pinion machine setting parameters. He used FEM to compare models generated for different pinion machine setting parameters. Castro [20] investigated the influence of mass temperature on gear scuffing and developed a new scuffing parameter for gears lubricated with mineral base oils. The scuffing parameter is based on the heat power intensity for mass temperature calculations. Mieth [21] presented a method for stress calculation on bevel gears with FEM influence vectors, in which these vectors allowed the consideration of the tooth geometry and gear-body constraint in the load distribution.

Wang [22] proposed a methodology for optimising the loaded contact pattern of spiral bevel and hypoid gears that solved the optimisation with a surrogate kriging-based model. This model considered the contact pattern under load, loaded transmission error, contact strength, and bending strength. The paper presented a numerical example that reduced the loaded transmission error by 30.3% by decreasing the maximum contact stress and root bending stress.

Marciniec [23] compared numerical methods to determine the contact pattern of Gleason-type bevel gears. He used a mathematical model of tooth contact analysis and FEM to simulate the load. Li [24] analysed the thermal characteristics of spur/helical gear transmission using FEM. He derived a calculation formula of the frictional heat flux and convective heat transfer coefficient which considers different surfaces of the gear tooth. His approach reveals the temperature distribution on the tooth flank and provides theoretical guidance for gear optimisation and anti-scuffing capability. Rong [25] proposed innovative digital twin modelling for loaded contact pattern-based grinding.

De Bechillion [26] investigated scuffing with an experiment by using twin-disc contact with an experimental methodology to assess the initiation of gear scuffing. Experiments were conducted with nitrided steel and synthetic oil. He identified oil-film thickness as the key parameter, also showing that a variety of operating conditions influence scuffing as well. Scuffing could also be triggered by rising bulk temperatures.

## 2. Scuffing in Worm Gears

According to ISO 14635-1, scuffing is defined as "a particularly severe form of damage to the gear-tooth surface in which seizure or welding together of areas of tooth surface occur, due to absence or breakdown of a lubricant film between the contacting tooth flanks of mating gears" [27]. Typical reasons for the occurrence of scuffing are high temperatures and high pressure, while scuffing also most likely occurs in the case of high surface velocities. Scuffing depends on various properties, such as gear materials, lubricants, the roughness of mating surfaces, sliding velocities, and load.

The analysis of these properties shows the following:

- In the case of a worm-gear pair, the typical material for a worm is hardened steel, while a worm gear is usually made of bronze. The reason behind this lies in the fact that this material pair has the highest resistance to the occurrence of scuffing even though bronze is more expensive compared to other steels. The dissimilarity in materials reduces friction and therefore increases scuffing resistance.
- The sliding velocity is very high in worm gears because they transfer motion through a 90° axis angle. This is a factor that significantly affects the occurrence of scuffing, and it cannot be influenced.
- When it comes to lubrication, it is recommended to lubricate worm gears in extreme operating environments, such as certain oils, with extreme-pressure (EP) additives, so as to increase resistance to the absence or breakdown of a lubricant film, therefore the occurrence of reducing scuffing.
- The roughness of mating surfaces directly influences the occurrence of scuffing; the rougher the surfaces, the higher the likelihood of occurrence of scuffing.

This analysis shows that scuffing is crucial for the design of worm gears, and it requires a significant amount of research to control the occurrence of scuffing and therefore increase scuffing load capacity. The idea of this paper is to investigate the influence of the contact pattern of a worm gear set on friction heat generation during meshing using FEM simulation and to check to what extent it can influence heat generation and, therefore, the occurrence of scuffing.

The contact pattern between a worm and worm gear in a standardly manufactured and properly mounted gear pair should be in the middle and across the worm-gear tooth. A worm gear is manufactured with a helix angle, and this means that the geometry of the worm-gear tooth differs at the input and output sides.

For the FEM simulation of a worm gear set, the referenced geometry according to ISO/TS 14521 was chosen [2]. Changing the pinion machine setting parameters can influence the contact pattern and therefore change the contact shape and the contact heat flux. It is then possible to find optimal machine setting parameters from the aspect of minimising the heat flux and, thus, increasing the efficiency of transmission. In this paper, direct-coupled field thermal–structural FEM simulation is used to investigate the difference of the contact heat flux at the contact pattern for standard-production machine settings and those that shift contact patterns for the same value into the inlet and outlet side of the worm-gear tooth. The results of this paper can be used by worm-gear designers to decrease the occurrence of scuffing by changing the contact pattern. This paper also shows the potential of contact patterns to increase load capacity.

## 3. Geometry

The geometry of the referenced gear according to ISO/TS 14521 [2] was used for the simulation. Table 1 lists the worm-gear pair's geometry, materials, and thermal data. Figure 1 represents the worm-gear set with the referenced geometry listed in Table 1. The structures of the worm and worm-gear sub-solids were designed to facilitate the generation of a uniform high-quality mesh for finite element analysis.

**Table 1.** Tooth geometry, materials, and thermal data of the worm-gear pair used.

| Parameter | Worm | Worm Gear |
|---|---|---|
| Flank shape | | ZI |
| Centre distance [mm] | | 100 |
| Axial module [mm] | | 4 |
| Pressure angle at normal section [°] | | 20 |
| Profile shift coefficient | | 0 |
| Mean lead angle of the worm [°] | | 12.5288 |
| Hand of gear | | right |
| Number of teeth | 2 | 41 |
| Reference operating diameter [mm] | 36 | 164 |
| Wheel width [mm] | 60 | 30 |
| Material | 16MnCr5 case hardening | CuSn12-G-GZ |

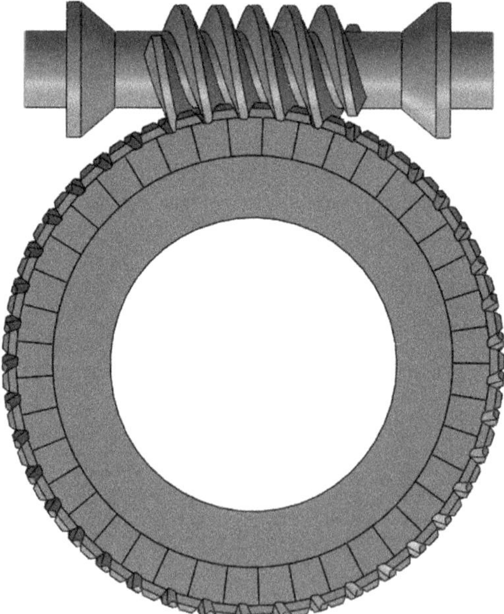

**Figure 1.** CAD model of worm-gear pair.

The whole gear pair was generated in SolidWorks and then transferred to Ansys 19.2 for further analysis. For a better understanding of the worm-gear contact pattern, a barrel-shaped tooth of the worm-gear tooth with 10 μm was introduced. The correct geometry of the worm and worm-gear teeth as well as the contact pattern was created using TRABI 8.0 software [27]. In order to generate correct geometry, special attention was given to the contact flanks. These flanks were created by first entering individual points generated in TRABI and then creating the curves based on those points that we used for the generation of surfaces, and by creating solids that we based on those surfaces (Figure 2).

**Figure 2.** Worm-gear tooth.

## 4. Contact Pattern

The gear-tooth contact pattern is the gear axial location used to achieve the desired position of the gear in order to provide an initial contact pattern to carry the design load of the gear set. The standard calculation for manufacturing worm gears yields a contact pattern with a calculated contact pattern in the middle. The authors chose to generate three models that could be used to compare the dependence of contact patterns on friction heat generation. The machine settings for the worm gear are given in Table 2.

**Table 2.** Machine settings for the radial milling cutter.

| Parameter | Value |
|---|---|
| Module [mm] | 3.994 |
| Pressure angle at normal section [°] | 19.8 |
| Mean diameter [mm] | 36 |
| Machine settings | |
| Axial displacement [mm] | 0 |
| Radial displacement [mm] | 0 |
| Pressure angle [°] | 0.0182 |

The typical method for manufacturing worm gears is hobbing using a hob or cutting tool. The cutting tool for worm gears is similar to the gear with which the worm gear will mate.

Figure 3 shows the contact pattern of a worm gear with the contact in the middle, as is the case with regular machine setting parameters.

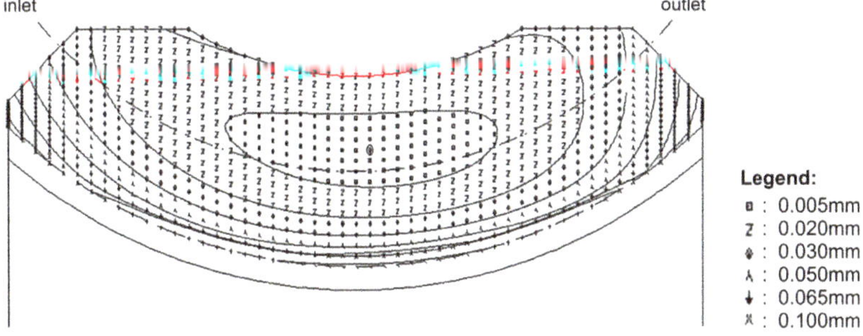


**Figure 3.** Contact pattern of the version in the middle.

Figure 4 shows the contact pattern at a worm gear that is shifted left to the outlet side, while Figure 5 shows the contact pattern at a worm gear that is shifted right to the inlet side. The difference is a machine setting pressure angle that is increased to 0.0518 and equals 0.07 in the case in Figure 4. In the case when the contact pattern is shifted to the inlet side, this parameter is decreased to −0.0518, as shown in Figure 5.

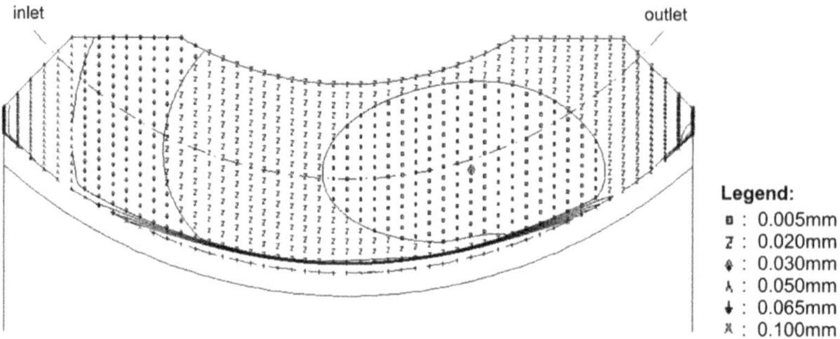


**Figure 4.** Contact pattern at the worm gear with the contact shifted to the outlet side.

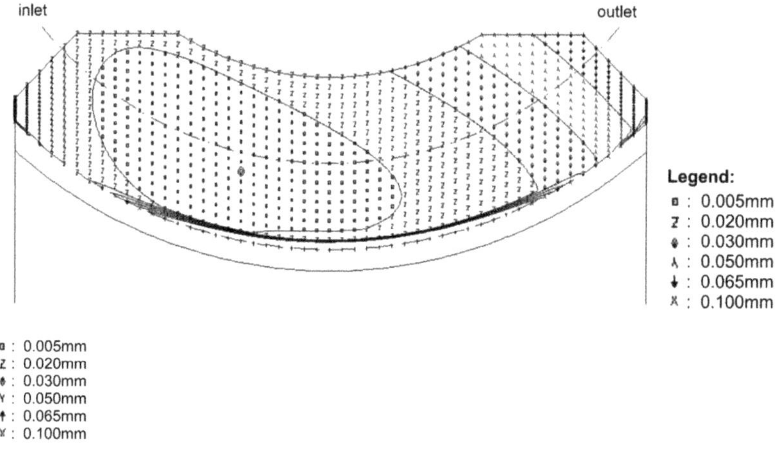


□ : 0.005mm
Z : 0.020mm
♦ : 0.030mm
Υ : 0.050mm
↑ : 0.065mm
Χ : 0.100mm

**Figure 5.** Contact pattern at the worm gear with the contact shifted to the inlet side.

## 5. Finite Element Modelling and Simulation

Direct-coupled transient thermal–structural analysis was used to assess the heat flux distribution of the worm-gear pair. The modelling and solving of the finite element model were conducted using ANSYS 19.2 software.

The rate of frictional dissipation in the contact elements was evaluated using the default ANSYS equation [28]:

$$q = FHTG \cdot \tau \cdot v, \tag{1}$$

where:

$FHTG$—the dissipation factor, which takes into account the part of friction energy that is converted into heat;

$\tau$—the equivalent stress that depends on the contact pressure and the friction coefficient;

$v$—the relative sliding velocity.

All frictional dissipated energy was considered to be converted into heat, and the distribution of friction-generated heat between the worm and the worm gear was considered equal.

The numerical analysis was carried out via three main steps according to the operating conditions listed in Table 3. The model's boundary conditions are defined in Figure 6. The first step was the introduction of the axial force load to establish proper contact between the worm and the wheel. The second step was to increase the speed of the worm drive to the operating speed of 1500 min$^{-1}$. The third step was used to assess the heat flux distribution in the operating conditions. The analysis time of the third step that corresponded to the actual operating conditions was 0.4 s, which enabled the gear pair to rotate for several mesh cycles and to achieve consistent results, showing dynamic changes in temperature and contact stress.

**Table 3.** Operating, material, and thermal data of the worm-gear pair used.

| Parameter | Worm | Worm Gear |
| --- | --- | --- |
| Power [kW] | 4.5 | |
| Speed of worm [$min^{-1}$] | 1500 | |
| Torque at worm gear [Nm] | 587.28 | |
| Mean tooth friction number | 0.0258 | |
| Heat transfer coefficient [(W/m2/K)] | 24,440 | |
| Young's modulus [Mpa] | 206,000 | 88,300 |
| Poisson's ratio | 0.300 | 0.350 |
| Specific heat [J/kgK] | 434 | 384 |
| Thermal conductivity [W/(m·K)] | 60.5 | 60.4 |
| Coefficient of thermal expansion [$C^{-1}$] | $1.2 \times 10^{-5}$ | $1.85 \times 10^{-5}$ |

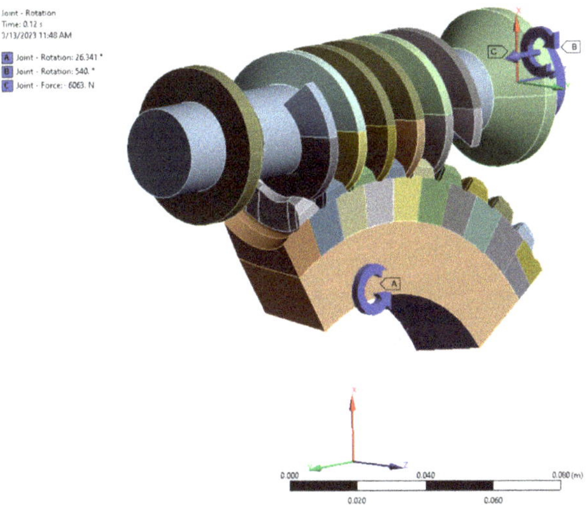


**Figure 6.** FEM setup for the directly coupled transient thermal–structural analysis.

The central rotation points of the pinion and gear were fixed to the inner side of the worm and wheel. The geometry of the worm and wheel was sliced to obtain simple bodies for optimising the mesh size and quality. The final mesh contained 24,703 elements with 27,926 nodes. The mesh of the worm-gear tooth is shown in Figure 7, and the contact surfaces of the worm and worm gear are shown in Figure 8. The mesh was generated with higher-order elements (SOLID226 [29]) in order to accurately capture the geometry of the worm and wheel. A mesh sensitivity test was performed, and mesh sizing of 1 mm on tooth flanks was adopted to ensure that contact pressure results did not differ by more than 5%.

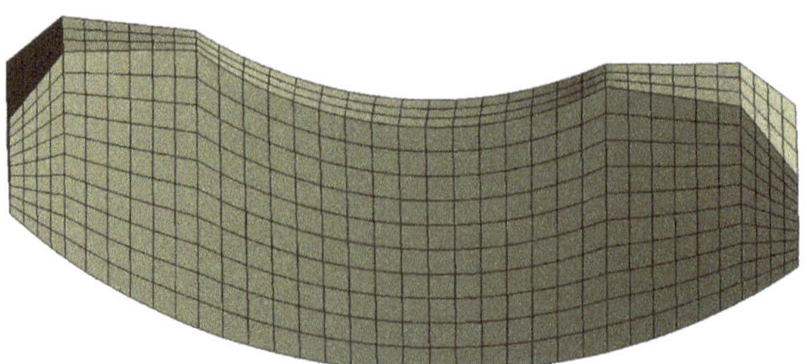

**Figure 7.** Mesh at the worm-gear tooth.

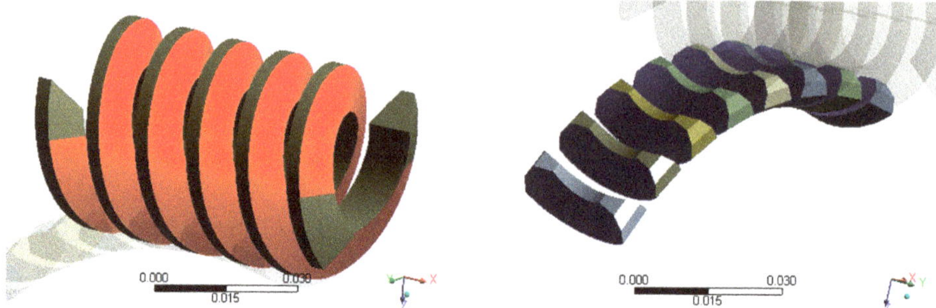

**Figure 8.** Contact surfaces of worm and worm gear.

The friction coefficient value necessary for the definition of contact pairs was determined based on ISO/TS 14521. The friction coefficient, i.e., mean tooth friction number, as defined in the standard was calculated as:

$$\mu_{zm} = \mu_{0T} \cdot Y_S \cdot Y_G \cdot Y_W \cdot Y_R = 0.0245, \tag{2}$$

In Equation (2), the size factor $Y_S$ takes into account the influence of the centre distance via the following formula:

$$Y_S = (100/a)^{0.5} = 1, \tag{3}$$

The geometry factor $Y_G$ considers the gear geometry's influence on the lubricating gap thickness. It depends on the characteristic value for mean lubrication gap width $h^* = 0.0692$ that is calculated according to ISO 14521.

$$Y_G = \left(0.07/h^*\right)^{0.5} = 1.006, \tag{4}$$

The material coefficient $Y_W$ takes into account the material of the wheel, and for the materials that are used in this paper, the coefficient was selected as $Y_W = 0.95$ from DIN EN 1982 [3]. The roughness factor $Y_R$ takes into account the influence of the surface roughness of the worm flank, and for $R_{a1} = 0.6$, it is calculated as follows:

$$Y_R = \sqrt[4]{R_{a1}/0.5} = 1.047, \tag{5}$$

The base friction number depends on the oil type and the material of the wheel, and for oil-bath lubrication and oil ISO-VG 220 [3] with synthetic oil based on Polyglycol, the following equation applies:

$$\mu_{0T} = 0.024 + 0.0032 \cdot \frac{1}{\left(v_{gm} + 0.1\right)^{1.71}} = 0.02449, \tag{6}$$

The sliding velocity on the mean circle (m/s) is calculated as follows:

$$v_{gm} = \frac{d_{m1} \cdot n_1}{19,098 \cdot \cos \gamma_m},$$

(7)

The heat transfer coefficient of the wheel (W/m$^2$/K) is calculated as follows:

$$\alpha_L = c_k \cdot (1940 + 15 \cdot n_1)$$

(8)

where $ck$ = 1 is the immersion factor for the wheel, and for the immersed wheel, a value of 1 was used (for oil-bath lubrication) and $n_1$ is speed of worm.

## 6. Results

Figure 9 shows the pressure distribution on the worm-gear teeth for all three cases. It shows that the pressure is in the centre for the case with the standard contact pattern (a) or slightly to the left (b), corresponding to the shift to the outlet side, or to the right (c), corresponding to the contact pattern shift to the inlet side.

Figure 10 shows the heat flux distribution on the worm-gear teeth with the standard contact pattern on two and three teeth. In this case, it is clear that the maximal heat flux is approximately in the middle of the gear teeth, as expected from the contact pressure results.

Figure 11 shows the heat flux distributions on the worm-gear teeth with the contact pattern shifted to the outlet side on two and three teeth. In this case, it is clear that the distribution of the heat flux as well as the maximal heat flux are on the right part of the gear teeth.

Figure 12 shows the heat flux distribution on the worm-gear teeth with the contact pattern shifted to the inlet side on two and three teeth. In this case, it is clear that the distribution of heat flux as well as the maximal heat flux are on the left part of the gear teeth.

Figure 13 shows a comparison of the average heat flux on the worm-gear teeth over the operation time of 0.4 sec. The average heat flux for the contact surface had the normal variant value of 21,888.3 W/m$^2$. For the worm-gear variant where the contact pattern was shifted to the inlet side, the average heat flux was 21,040.1 W/m$^2$, and for the variant where the contact pattern was shifted to the outlet side, the average heat flux was 23,440.3 W/m$^2$. This means that when compared to the standard contact pattern, the average heat flux for the worm gear with the contact pattern shifted to the inlet side was 3.76% lower, while it was 7.09% higher for the contact pattern that was shifted to the outlet side.

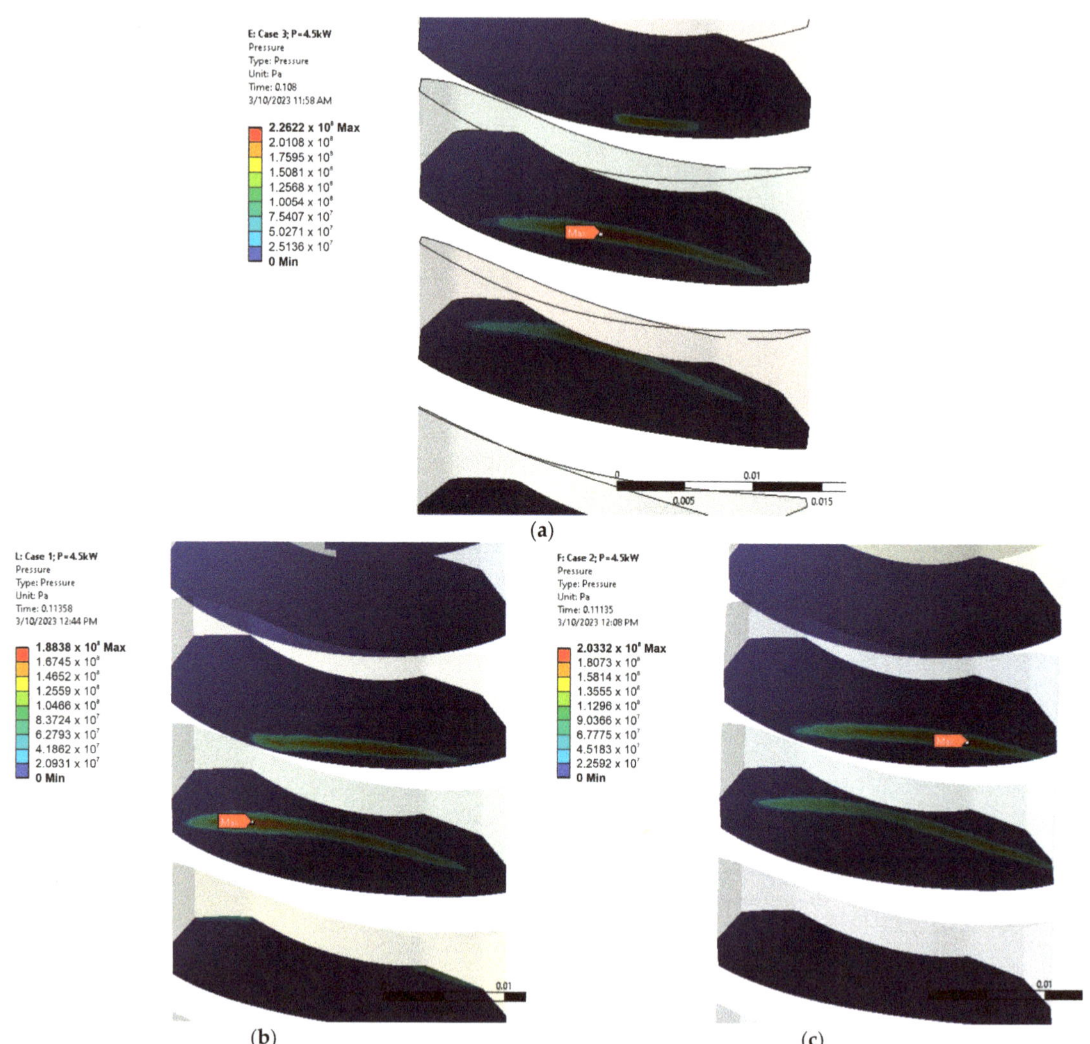

**Figure 9.** Pressure distribution in the worm gear: (**a**) normal contact pattern; (**b**) contact pattern on the outlet side; (**c**) contact pattern on the inlet side.

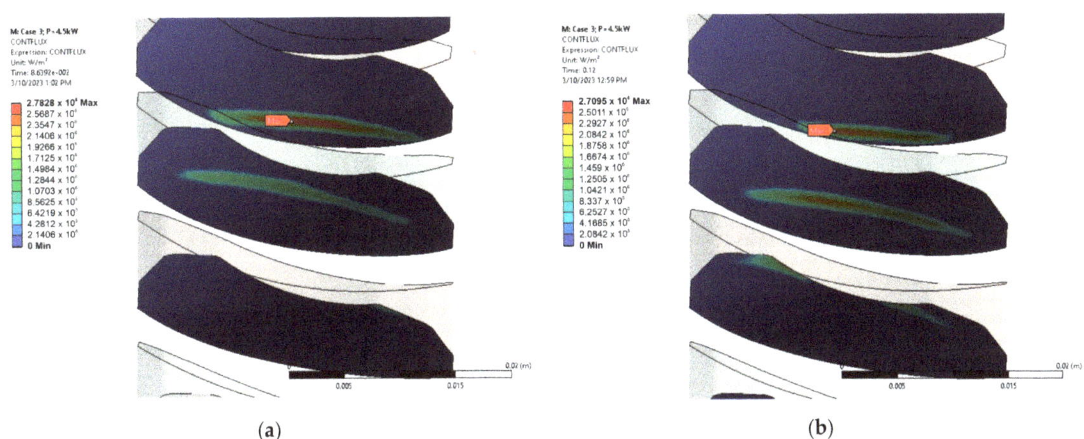

(a)           (b)

**Figure 10.** Heat flux distributions for the standard contact pattern: (**a**) contact on two teeth, (**b**) contact on three teeth.

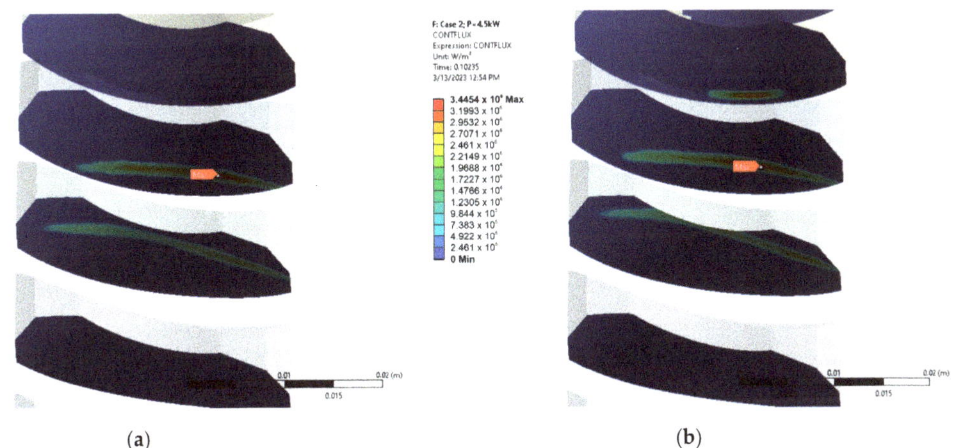

(a)           (b)

**Figure 11.** Heat flux distributions for the contact pattern shifted to the outlet side: (**a**) contact on two teeth, (**b**) contact on three teeth.

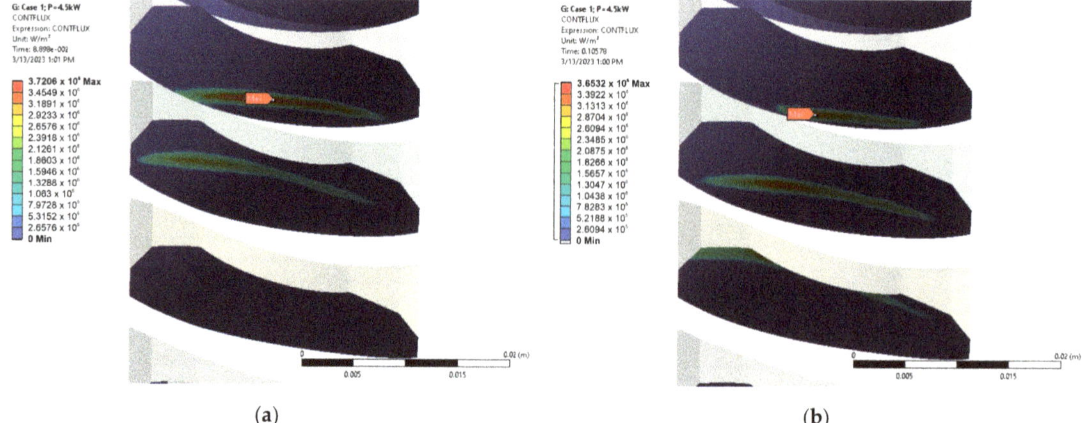

(**a**)                                                                          (**b**)

**Figure 12.** Heat flux distribution for the contact pattern shifted to the inlet side: (**a**) contact on two teeth, (**b**) contact on three teeth.

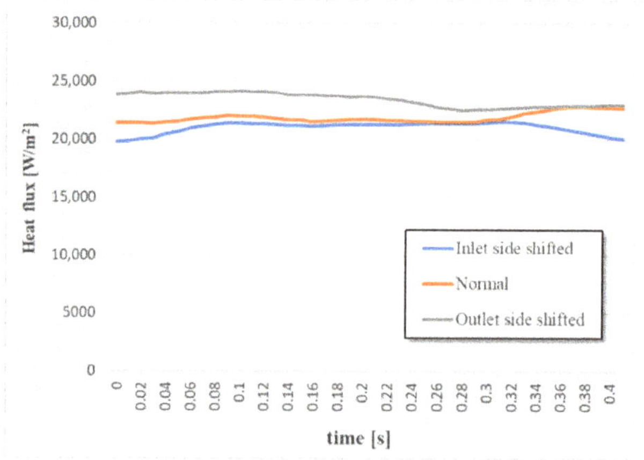



**Figure 13.** Differences in heat flux for three variants.

## 7. Conclusions

This paper investigated the influence of the contact pattern of a worm-gear set on the friction heat generated in the contact between the worm and the worm-gear teeth. Using FEM simulation, three cases of worm-gear pairs were investigated with different pinion machine setting parameters. The referent variant was the normal contact pattern according to the standard, with two additional variants in which the contact pattern was shifted either to the inlet or the outlet side of the worm-gear tooth.

The results of the FEM analysis confirmed the position of the contact pattern on the worm-gear tooth for all three cases. The average heat flux had the highest values when the contact pattern was shifted to the outlet side, and it was 7.09% higher than for the standard contact pattern. The worm-gear tooth with the contact pattern shifted to the inlet side had a value of the average heat flux that was 3.76% lower than the case with the standard contact pattern.

The results indicate that shifting the contact pattern to the inlet side generates slightly less heat during operation. Therefore, such a gear pair will be less susceptible to the occurrence of scuffing and have a higher load capacity. On the other hand, shifting the contact pattern to the outlet side will result in opposite effects—lower efficiency and load capacity.

Further research should go in the direction of experimental verification of the results that are presented in this paper. It would be beneficial to find an efficient way to collect thermal data during the operation of experiments since the thermal distribution on the surfaces in contact is very important for the occurrence of scuffing.

**Author Contributions:** Conceptualisation, A.M. and M.B.; methodology, A.M.; software, M.B. and N.V.; validation, M.P. and D.R.; formal analysis, M.S.; investigation, A.M.; writing—original draft preparation, A.M.; writing—review and editing, M.B. and M.S.; visualisation, M.P. and D.R.; supervision, M.S.; project administration, A.M. All authors have read and agreed to the published version of the manuscript.

**Funding:** This research was financially supported by the Ministry of Science, Technological Development, and Innovation of the Republic of Serbia (Contract No. 451-03-47/2023-01/200109).

**Institutional Review Board Statement:** Not applicable.

**Informed Consent Statement:** Not applicable.

**Data Availability Statement:** Data are contained within the article.

**Conflicts of Interest:** The authors declare no conflict of interest.

# References

1. Niemann, G.; Winter, H. *Machine Elements Volume 3: Helical, Bevel, Worm, Chain, Belt, Friction Gears, Clutches, Brakes, Freewheels*; 2., völlig neu bearbeitete Auflage; Springer: Berlin/Heidelberg, Germany, 2004; ISBN 978-3-642-62101-7.
2. *ISO/TS 14521:2020-04*; Gears-Calculation of Load Capacity of Worm Gears. International Organization for Standardization: Geneva, Switzerland, 2020.
3. *DIN 3996:2019-09*; Calculation of Load Capacity of Cylindrical Worm Gear Pairs with Rectangular Crossing Axes. Deutsches Institut Fur Normung E.V.: Berlin, Germany, 2019.
4. Höhn, B.R.; Steingröver, K.; Lutz, M. Determination and Optimization of the Contact Pattern of Worm Gears. *Gear Technol.* **2003**, *20*, 12–17.
5. Simon, V. Computer Aided Loaded Tooth Contact Analysis in Cylindrical Worm Gears. *ASME J. Mech.* **2004**, *127*, 973–981. [CrossRef]
6. Sharif, K.J.; Evans, H.P.; Snidle, R.W. Prediction of the wear pattern in worm gears. *Wear* **2006**, *261*, 666–673. [CrossRef]
7. Paschold, C.; Sedlmair, M.; Lohner, T.; Stahl, K. Efficiency and heat balance calculation of worm gears. *Forsch Ingenieurwes* **2020**, *84*, 115–125. [CrossRef]
8. Roth, F.; Hein, M.; Stahl, K. Scuffing load capacity calculation of worm gears. *Forsch Ingenieurwes* **2021**, *86*, 503–511. [CrossRef]
9. Stahl, K.; Höhn, B.-R.; Hermes, J.; Monzn, A. Pitting Resistance of Worm Gears: Advanced Model for Contact Pattern of Any Size, Position, Flank Type. *Gear Technol.* **2012**, *10*, 44–49.
10. Hiltcher, Y.; Guing, M.; Vaujany, J.P. Numerical simulation and optimisation of worm gear cutting. *Mech. Mach. Theory* **2006**, *41*, 1090–1110. [CrossRef]
11. Daubach, K.; Oehler, M.; Sauer, B. Wear simulation of worm gears based on an energetic approach. *Forsch Ingenieurwes* **2021**, *85*, 367–377. [CrossRef]
12. Tošić, M.; Larsson, R.; Lohner, T. Thermal Effects in Slender EHL Contacts. *Lubricants* **2022**, *10*, 89. [CrossRef]
13. Tošić, M.; Larsson, R.; Stahl, K.; Lohner, T. Thermal Elastohydrodynamic Analysis of a Worm Gear. *Machines* **2023**, *11*, 89. [CrossRef]
14. Chernets, M.V. Prediction Method of Contact Pressures, Wear and Life of Worm Gears with Archimedean and Involute Worm, Taking Tooth Correction into Account. *J. Frict. Wear* **2019**, *40*, 342–348. [CrossRef]
15. Jbily, D.; Guingand, M.; de Vaujany, J.-P. A wear model for worm gear. *Proc. Inst. Mech. Eng. Part C J. Mech. Eng. Sci.* **2015**, *230*, 1290–1302. [CrossRef]
16. Oehler, M.; Magyar, B.; Sauer, B. Ein neuer, normungsfähiger Berechnungsansatz für den Wirkungsgrad von Schneckengetrieben. *Forsch Ingenieurwes* **2017**, *81*, 145–151. [CrossRef]
17. Miltenović, A.; Tica, M.; Banić, M.; Miltenović, Đ. Prediction of Temperature Distribution in the Worm Gear Meshing. *Facta Univ. Ser. Mech. Eng.* **2020**, *18*, 329–339. [CrossRef]
18. Li, S.; Kahraman, A. A scuffing model for spur gear contacts. *Mech. Mach. Theory* **2021**, *156*, 104161. [CrossRef]
19. Yanzhong, W.; Wen, T.; Yanyan, C.; Yanyan, C.; Wang, T.J.; Li, G. Investigation into the meshing friction heat generation and transient thermal characteristics of spiral bevel gears. *Appl. Therm. Eng.* **2017**, *119*, 245–253.
20. Castro, J.; Seabra, J. Influence of mass temperature on gear scuffing. *Tribol. Int.* **2018**, *119*, 27–37. [CrossRef]

21. Mieth, F.; Ulrich, C.; Schlecht, B. Stress calculation on bevel gears with FEM influence vectors. *Forsch Ingenieurwes* **2022**, *86*, 491–501. [CrossRef]
22. Wang, Q.; Zhou, C.; Gui, L.; Fan, Z. Optimization of the loaded contact pattern of spiral bevel and hypoid gears based on a kriging model. *Mech. Mach. Theory* **2018**, *122*, 432–449. [CrossRef]
23. Marciniec, A.; Pacana, J.; Pisula, J.M.; Fudali, P. Comparative analysis of numerical methods for the determination of contact pattern of spiral bevel gears. *Aircr. Eng. Aerosp. Technol.* **2018**, *90*, 359–367. [CrossRef]
24. Li, W.; Zhai, P.; Ding, L. Analysis of Thermal Characteristic of Spur/Helical Gear Transmission. *ASME J. Thermal Sci. Eng. Appl.* **2019**, *11*, 021003. [CrossRef]
25. Rong, K.; Ding, H.; Kong, X.; Huang, R.; Tang, J. Digital twin modeling for loaded contact pattern-based grinding of spiral bevel gears. *Adv. Eng. Inform.* **2021**, *49*, 101305. [CrossRef]
26. De Bechillon, N.G.; Touret, T.; Cavoret, J.; Changenet, C.; Ville, F.; Ghribi, D. A new experimental methodology to assess gear scuffing initiation. *Tribol. Mater. Surf. Interfaces* **2022**, *16*, 245–255. [CrossRef]
27. *ISO 14635-1:2023*; Gears—FZG Test Procedures—Part 1: FZG Test Method A/8, 3/90 for Relative Scuffing Load-Carrying Capacity of Oils. International Organization for Standardization: Geneva, Switzerland, 2023.
28. *Benutzerbeschreibung und Programmdokumentation zum Programmpaket TRABI/TBV, LMGK*; Ruhr-University Bochum: Bochum, Germany, 2006.
29. *ANSYS Mechanical APDL Contact Technology Guide*; Realise 15; ANSYS Inc.: Canonsburg, PA, USA, 2013.

*applied sciences*

*Article*

# Numerical and Experimental Analysis of the Oil Flow in a Planetary Gearbox

Marco Nicola Mastrone [1], Lucas Hildebrand [2], Constantin Paschold [2], Thomas Lohner [2], Karsten Stahl [2] and Franco Concli [1,*]

[1]   Faculty of Science and Technology, Free University of Bolzano, Piazza Università 5, 39100 Bolzano, Italy
[2]   Gear Research Center (FZG), Department of Mechanical Engineering, School of Engineering and Design, Technical University of Munich, Boltzmannstraße 15, 85748 Garching, Germany
*   Correspondence: franco.concli@unibz.it

**Abstract:** The circular layout and the kinematics of planetary gearboxes result in characteristic oil flow phenomena. The goal of this paper is to apply a new remeshing strategy, based on the finite volume method, on the numerical analysis of a planetary gearbox and its evaluation of results as well as its validation. The numerical results are compared with experimental data acquired on the underlying test rig with high-speed camera recordings. By use of a transparent housing cover, the optical access in the front region of the gearbox is enabled. Different speeds of the planet carrier and immersion depths are considered. A proper domain partitioning and a specifically suited mesh-handling strategy provide a highly efficient numerical model. The open-source software OpenFOAM® is used.

**Keywords:** CFD; planetary gearboxes; high-speed camera; lubrication; oil flow

## 1. Introduction

Planetary gearboxes are widely used in mechanical transmissions with great advantages, such as high transmissible torque, power density, high reliability, and efficiency. These gearboxes are composed of four main components: the sun gear, the ring gear, the planet gears, and the planet carrier. The sun is an external gear and is coaxial with the ring gear, which, on the contrary, is an internal gear. The planets are external gears that engage with both the sun and the ring gears. Different arrangements of the sun, the planets, the planet carrier, and the ring gear are possible due to the kinematic degrees of freedom.

Planetary gearboxes are used in several application areas, especially in industrial automation systems, wind power industry, and industrial gear units. For example, they are used in dynamic applications in robotics, where factors such as low backlash and a balanced load distribution are highly valued so to ensure a longer system lifetime and positioning accuracy. Typical applications include machine tools and self-propelled machines, packaging systems, conveyor belts, printing presses, and any context where extremely precise positioning control is required. Compared with other types of geared transmissions, planetary gearboxes allow to design a more compact system, a feature that is always appreciated by machine designers.

In geared transmissions, it is crucial to ensure the proper lubrication of all components, especially when they operate at high speeds or must withstand heavy loads. In this work, a combined experimental–numerical approach is used to investigate the oil phenomena that occur in the FZG internal gear test rig [1,2]. A special transparent housing cover was designed to make the front part of the gearbox visible. This allowed to capture the oil flow by high-speed camera recordings. To analyze the oil distribution, a CFD model was developed. A new mesh-handling technique was applied to build a time efficient simulation model. The CFD results show a high comparability with experimental results for a wide range of operation conditions. The potential of CFD for predicting oil flow is shown.

**Citation:** Mastrone, M.N.; Hildebrand, L.; Paschold, C.; Lohner, T.; Stahl, K.; Concli, F. Numerical and Experimental Analysis of the Oil Flow in a Planetary Gearbox. *Appl. Sci.* **2023**, *13*, 1014. https://doi.org/10.3390/app13021014

Academic Editor: Filippo Berto

Received: 21 December 2022
Revised: 9 January 2023
Accepted: 10 January 2023
Published: 11 January 2023

*Literature Review*

The first studies on the no-load power losses arising from the interaction of the moving components with the lubricant in planetary gears date back to Gold et al. [3]. They performed theoretical and experimental studies aimed at studying a two-stage planetary gearbox partially immersed in oil. Further analyses by Kettler [4] concentrated on the operating temperature and the heat balance in dip-lubricated planetary gearboxes. The geometrical parameters of the gearbox such as center distance, planet carrier geometry, and the number of planets, as well as the physical properties of the system such as lubricant parameters and immersion depth, were considered to derive analytical equations for the determination of no-load power losses. Subsequent analyses by Da Costa [5] dealt with the impact of different types of lubricant, namely mineral and synthetic oil, on the no-load power losses for dip lubrication. Studies with injection lubrication were conducted by Höhn et al. [6] and Schudy [1], who built the FZG internal gear test rig to investigate the flank load carrying capacity of internal gears, and the influence of circumferential speeds, material combinations, surface finishing, and lubricants. Suggestions for lubrication conditions, surface finishing, and drive direction were provided. De Gevigney et al. [7] performed an experimental campaign on a planetary gearbox aimed at studying the impact of injection volume rate, lubricant temperature, and rotational speed on efficiency. Recent experiments by Boni et al. [8,9] focused on the detailed analysis of the oil distribution in a dip-lubricated planetary gearbox and its efficiency. They used a front housing made from acryl glass that allowed observation of the oil level and its flow during operation. Various observations were made by considering different rotational speeds, lubricant temperatures, and filling levels. As the planet carrier starts to rotate, the oil tends to distribute around the circumference of the covering box creating a dynamic peripheral oil ring. The authors considered the thickness of the oil ring for the estimation of the churning power losses. This phenomenon has not been observed for other gearbox designs, meaning that it is a peculiar characteristic of planetary gearboxes.

Due to the difficult visibility of internal gearbox components, it is not trivial to have a deep understanding of the lubrication mechanisms, especially for systems characterized by a compact design as planetary gearboxes. However, recent developments in computer science made it possible to use numerical approaches to study the specific problem of interest. In the case of gearbox lubrication, Computational Fluid Dynamics (CFD) can offer substantial benefits in the prediction of the oil flow inside the gearbox. Indeed, in the post-processing phase, it is possible to analyze the oil distribution inside the considered gearbox and to derive information on the lubrication behavior of the system. However, due to the complex kinematics of planetary gearboxes and the high computational resources needed, the numerical studies available in the literature are limited. Moreover, sometimes significant modeling simplifications have been adopted, making it difficult to assess the models' reliability concerning the oil flow predictions. A recent exhaustive literature review on CFD methods applied to the study of gearbox lubrication and efficiency can be found in Maccioni et al. [10]. In the following, the sole CFD studies relative to epicyclic gearboxes are reported.

Bianchini et al. [11,12] exploited the Multiple Reference Frame (MRF) approach to model a planetary gearbox for aeronautic applications. They increased the gear interspacing, i.e., the meshing region between the sun and the planets, and between the planets and the ring gear, was not modelled. In this way, the squeezing effects could not be simulated and the related axial phenomena occurring in the mating zone were neglected. Moreover, only single-phase simulations were performed. The adopted MRF technique is limited to a regime solution of an unsteady problem without information on the transient behavior. Despite low computational effort, the approach lacks an accurate description of multiphase operating conditions and oil flow phenomena.

More advanced numerical techniques, such as the sliding mesh approach and the dynamic remeshing, are suitable for simulating gear rotation and interaction. FZG test rigs are used for experimental studies on efficiency and lubrication [13]. Concli et al. [14,15]

implemented a CFD model of a planetary gearbox studying the no-load power losses for different oil temperatures, immersion depths, and planet carrier rotational speeds. The results were validated with experimental data in terms of power losses. Further studies by Concli et al. [16–18] introduced an innovative remeshing procedure in the opensource CFD code OpenFOAM® [19] for the simulation of the gear meshing. The strategy is based on the computation of extruded numerical grids connected through numerical interfaces to link the non-conformal nodes. When the quality of the mesh decreases, a new mesh is computed for the new gear position. The results are interpolated and the simulation proceeds. The virtual models could predict the efficiency of the system accurately. This algorithm requires the generation of a new mesh whenever bad-quality elements originate. This means that hundreds of meshes may be needed to complete the simulation.

Cho et al. [20] implemented a CFD model based on the overset mesh approach (also called the overlapping grids method, or chimera framework) [21,22] of a planetary gearbox. Filling levels of 30%, 50%, and 70% of the gearbox volume were analyzed. Owing to the numerical analysis, the system design could be optimized and the thermal behavior could be determined. The authors suggest using high filling levels to increase the cooling effects despite the higher churning losses. The commercial software STAR-CCM+ [23] was used.

Liu et al. [2] investigated the oil distribution of the FZG internal gear test rig, which is also the object of this study, for different oil properties and planet carrier speeds in detail. The finite volume-based CFD implementation included a dynamic remeshing approach to describe the rotation and engagement of the gears. They performed multiphase simulations by applying the Volume of Fluid (VoF) method. The roller bearings of the planet gear shafts were also modeled in the numerical simulation, but their kinematics were described with a pure rotation around the central axis of the planetary gearbox and not with a roto-translation, which resulted in doubling the computational time. They performed a mesh sensitivity analysis with respect to loss torque and oil distribution to obtain an efficient yet accurate modeling approach. The high level of detail of the developed CFD model allowed for the extraction of information on the oil flow inside the gearbox. Liu et al. [2] indicate that with a rather low filling level, it is possible to supply enough oil to the different machine elements. The commercial software ANSYS FLUENT [24] was used. The simulations were parallelized on 80 cores of a high-performing computer cluster.

As it can be noticed from the literature review on planetary gearboxes, the experimental studies are limited, and the numerical analyses mostly lack validation, particularly concerning the validation of the oil flow, since the internal part of the gearbox could not be visibly accessed. Moreover, significant computational resources were required to run such simulations. To increase confidence in simulation results, validation by experimental investigations in terms of oil flow is fundamental. Furthermore, efficient remeshing strategies can reduce the computational effort required for finite volume-based CFD simulations. This work fits into these objectives, providing side-by-side comparisons of high speed camera recordings and CFD results of the oil distribution. The system under consideration is the same investigated by Liu et al. [2]. As a new computationally efficient mesh-handling approach is introduced, the implemented numerical model is ideal to analyze the considered gearbox and can be run on "normal" computers without the need for high-performance computing clusters.

## 2. Materials and Methods

The FZG internal gear test rig is the subject of the investigation and served to validate the simulation results. In the following, a detailed description of the system and the operating conditions is given.

### 2.1. Object of Investigation

The object of investigation is the FZG internal gear test rig presented in Figure 1 (left). The test rig is composed of a planet carrier, three planet gears, and two ring gears, namely, a test ring gear and a drive ring gear. The planet carrier is driven by a continuously variable electric motor via a shiftable countershaft transmission. The three stepped planet gears of the test rig are arranged uniformly across the circumference and are engaged with both the test ring gear and the drive ring gear. The stages of the gearings exhibit the same gear ratio. The drive ring gear is float-mounted by spring elements which brace peripheral forces. Axial forces are supported by thrust washers mounted in the housing.

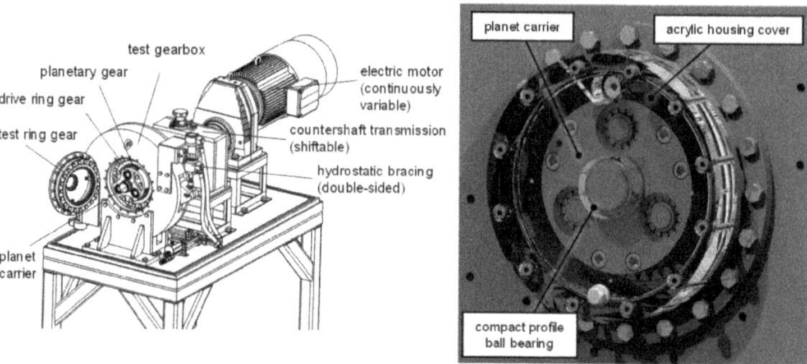


**Figure 1.** FZG internal gear test rig (**left**) and rendering of the modified FZG internal gear test rig with transparent housing cover (**right**) [25].

The FZG internal gear test rig was designed to operate with oil injection lubrication. Thereby, temperature-controlled oil is injected evenly around the circumference from an oil supply unit via two separate oil distributor pipes, one at the front side and one at the backside of the test rig. The test rig was modified to optically access the internals of the test rig and to realize an oil dip-lubricated operation. Therefore, the front facing housing cover was changed by a modified acrylic exchange part. Instead of the original cylinder roller bearing mounted to the planet carrier, a ball bearing with compact profile was mounted to improve visibility. The front-facing distributor pipe was removed to further improve visibility and realize an oil dip lubrication. In combination with a throttled oil backflow by an adjustable valve, a quasi-stationary oil level is maintained. A controlled oil sump temperature is achieved by a constant oil circulation and flow heaters. Figure 1 (right) shows a rendering of the test rig configuration used for the investigations in this study.

The gear geometry data are reported in Table 1. The mineral oil FVA3A (ISO VG 100) [26] was used as lubricant. The oil sump temperature was kept constant at 40 °C. In Table 2, the physical properties of the oil are reported. The camera used for the recordings is a Photron FASTCAM Mini AX200. A light source was exploited to produce high-intensity light directed towards the front cover.

**Table 1.** Gear geometry.

|  | Unit | Ring gear | Planet Gear |
|---|---|---|---|
| Center distance ($a$) | mm | 59 | |
| Normal module ($m_n$) | mm | 4.5 | |
| Number of teeth ($z$) | - | 42 | 16 |
| Face width ($b$) | mm | 16 | 14 |
| Tip diameter ($d_a$) | mm | 185.0 | 82.5 |
| Pressure angle ($\alpha_n$) | ° | 20 | |
| Helix angle ($\beta$) | ° | 0 | |
| Profile shift coeff. ($x$) | - | 0.1817 | 0.2962 |

**Table 2.** Lubricant properties FVA3A.

|  | Unit | Value |
|---|---|---|
| Kinematic viscosity (40 °C) | mm$^2$/s | 95 |
| Kinematic viscosity (100 °C) | mm$^2$/s | 10.7 |
| Density (40 °C) | kg/m$^3$ | 864 |

## 2.2. Operating Conditions

Following Liu et al. [2], planet carrier speeds ranging from 81 rpm to 324 rpm were considered in the experimental and numerical investigations so that the results could be compared with each other. Two oil filling levels were considered: a low one corresponding to 3 times the normal module with respect to the ring gears, as investigated by Liu et al. [2], and a high one corresponding to the centerline level (as a typical operating condition of planetary gearboxes, which was previously investigated numerically by Cho et al. [20] and experimentally by Boni et al. [8]). In Table 3, the analyzed operating conditions are summarized.

**Table 3.** Operating conditions.

| Simulation Name | Lubricant Filling Level | Planet Carrier Speed $n_t$ in rpm | Planet Gears Speed $n_{p,abs}$ in rpm | Tangential Speed at Pitch Circle $v_t$ in m/s |
|---|---|---|---|---|
| SIM-id1v1 |  | 81 | 132 | 0.8 |
| SIM-id1v2 | 3·m$_n$ | 162 | 263 | 1.6 |
| SIM-id1v3 |  | 324 | 526 | 3.2 |
| SIM-id2v1 |  | 81 | 132 | 0.8 |
| SIM-id2v2 | centerline | 162 | 263 | 1.6 |
| SIM-id2v3 |  | 324 | 526 | 3.2 |

"SIM" stands for simulation, "id" for immersion depth, and "v" for velocity. As already mentioned, two immersion depths and three rotational velocities were analyzed. Therefore, "id" has values of 1 and 2, while "v" has values of 1, 2, and 3.

## 2.3. Mathematical Description and Governing Equations

The CFD code used to perform the simulations is based on OpenFOAM® v7 [19], a Finite Volume Method (FVM) [27] software. The complexity of the two-phase oil–air splash lubrication in the planetary gearbox was modelled with the Volume of Fluid (VOF) model [28]. The sum of the two fluids volume ratios ($\alpha_{oil}$ and $\alpha_{air}$) in each control volume is equal to 1:

$$\alpha_{oil} + \alpha_{air} = 1 \tag{1}$$

When $\alpha_{oil} = 1$, the control unit is full of oil; when $\alpha_{oil} = 0$, the control unit is filled with air; when $0 < \alpha_{oil} < 1$, an oil–air interface is contained in the control unit. The solution of the continuity equation of the volume fraction allows to trace the oil–air interface.

$$\frac{\partial \alpha}{\partial t} + \nabla(\alpha u) = 0 \tag{2}$$

The Multidimensional Universal Limiter with Explicit Solution (MULES) [29] algorithm was used to achieve a better boundedness of the volume fraction field. This is achieved by adding a dummy velocity field ($u_c$) acting perpendicular to the interface to the conservation equation of the $\alpha$ field:

$$\frac{\partial \alpha}{\partial t} + \nabla(\alpha u) + \nabla(u_c \alpha(1 - \alpha)) = 0 \tag{3}$$

The generic property $\phi$ (as viscosity and density) of the equivalent fluid is calculated as:

$$\phi_{eq} = \alpha_{oil}\,\phi_{oil} + (1 - \alpha_{oil})\,\phi_{air} \tag{4}$$

The fluid-dynamic problem requires solving of the continuity and the momentum equations for the oil–air mixture:

$$\frac{\partial \rho}{\partial t} + \nabla(\rho u) = 0 \tag{5}$$

$$\frac{\partial(\rho u)}{\partial t} + \nabla(\rho uu) = -\nabla p + \nabla\left[\mu\left(\nabla u + \nabla u^T\right)\right] + \rho g + F \tag{6}$$

where $\rho$ is the density, $u$ is the fluid velocity, $\mu$ is the viscosity, $g$ is the gravity acceleration, and $F$ represents the vector of external forces. The weighted average properties are used so that the two conservation equations are associated with the continuity equation of the volume fraction.

### 2.4. CFD Modeling of the Oil Flow

In this section, the implementation of the CFD model in OpenFOAM® is discussed. A detailed description of the meshing procedure and of an innovative mesh-handling strategy for low computational effort is explained.

### 2.4.1. Geometrical Considerations

The CFD model was derived starting from the CAD model by considering the areas where the lubricant can be present. Therefore, different from the structural analysis, the negative model of the solid parts was considered. Small-scale features such as screws, springs, and seals were neglected as they are not expected to have a noteworthy impact on the oil distribution. Further cleaning of the geometry included the removal of chamfers, edges, and small clearances. The final geometry consists of two ring gears, three planets, and the planet carrier. To avoid numerical singularities in the gear meshing region, the ring gears were scaled to 102% of their nominal size (in the radial direction only). This preliminary action is required for all finite volume-based simulations of intersecting objects and is not expected to significantly influence the main oil flow in the gearbox.

### 2.4.2. Meshing Approach

As the investigated planetary gearbox is composed of spur gears, it is possible to exploit an extruded mesh approach. Extruded meshes are preferable over tetrahedral meshes, since they exhibit higher quality, lower memory usage, and faster calculation times. In a first step, the domain was decomposed in three subdomains which were meshed separately. The introduction of AMIs (Arbitrary Mesh Interfaces) [30] enabled the connection of adjacent non-conformal mesh domains. In this way, AMIs ensure the continuity of the field variables across the mesh discontinuities. In Figure 2, the three subdomains and the assembled domain are shown.

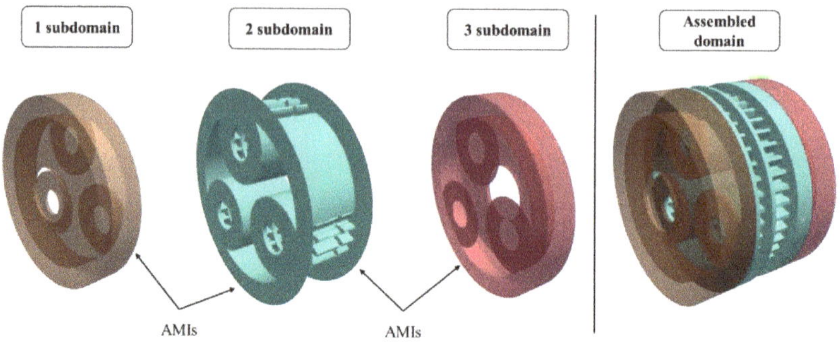


**Figure 2.** Domain decomposition: the AMIs enable the numerical connection among the 3 subdomains.

The first and the third subdomains are composed of the external parts of the gearbox and include a portion of the housing, the planet carrier, and the planet shafts. Their geometry is relatively simple, and an extruded mesh can be obtained. On the other side, the second subdomain contains the planets, the ring gears, and the remaining parts of the housing and of the planet carrier. The geometry of this subdomain is much more complex and, to generate a swept mesh, it is necessary to create axial partitions in correspondence with each tooth flank. Additional partitions with the shape of the planets, the ring gears, and the planet carrier profiles were created. Once all the necessary partitions were implemented, an extruded mesh was obtained for the second subdomain. In Figure 3, the longitudinal and the cross views of the assembled mesh are reported.

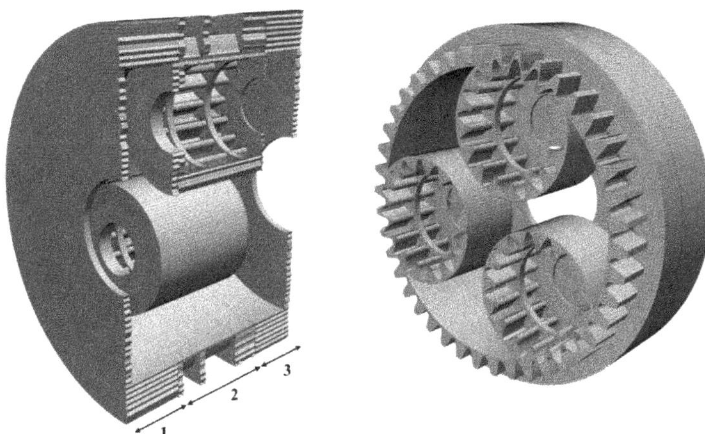


**Figure 3.** Longitudinal and cross views of the assembled mesh. A full extruded mesh is obtained.

The final mesh is composed of about 0.65 M prismatic elements. A grid sensitivity analysis of the CFD model was performed. The element number was varied from 0.37 M to 0.78 M. The maximum deviation of the power loss was only about 4%, even if the oil flow could be resolved better with increasing the element number. The grid with 0.65 M elements was chosen as a good compromise between oil flow resolution and computing time. The element size in the computational domain was of 1 mm. The analysis performed by Liu et al. [2] on the same system was composed of about 0.73 M tetrahedral elements, thus resulting in a comparable number of elements, but in a different cell type. The quality parameters of the initial and the displaced mesh at the end of the imposed motion are listed in Table 4. As can be seen, the max non-orthogonality (angle between the line connecting two cell centers and the normal of their common face) was kept below 70 and the maximum skewness (distance between the intersection of the line connecting two cell centers with their common face and the center of that face) was kept below 4. The geometry and the mesh computation phase was handled via Python [31] scripts and Salome [32].

**Table 4.** Quality parameters of the mesh.

| Max Non-Orthogonality [°] | | Max. Skewness [-] | |
|---|---|---|---|
| Initial mesh | Displaced mesh | Initial mesh | Displaced mesh |
| 50.7 | 66.1 | 1.6 | 2.8 |

### 2.4.3. Mesh-Handling Approach

The simulation of the considered gearbox requires the introduction of a dynamic mesh approach. Indeed, the boundary motion causes the deformation of the mesh, which needs to be replaced with a new valid one after every few simulated angular positions. The Local Remeshing Approach (LRA), available in some commercial software, can handle the boundary motion with deformable elements that stretch every timestep. The distorted elements that arise from the boundary node motion are automatically reconstructed by the code based on the imposed quality parameters. Despite its effectiveness, this method requires a sufficiently low timestep to ensure a stable remeshing process. This is mainly because each local remeshing iteration can lead to elements having a smaller size with respect to the original ones. Thus, the lower computational efficiency is related to an increase in the number of elements and the consequent decrease in the allowable timestep to keep the simulation stable. In the current work, an innovative remeshing approach was implemented, namely, an adapted version of the Global Remeshing Approach with Mesh Clustering (GRA$^{MC}$), which was already used by the authors to study spur, helical, and bevel gear pairs [33]. The main idea behind the GRA$^{MC}$ is that it is possible to compute a mesh set that covers one engagement and reuse it recursively during the simulation. In this way, the computational effort associated with the remeshing process is drastically reduced since the grids in the predefined wheel positions are immediately available for the mesh-to-mesh interpolation of the results. Moreover, this approach allows the user to have a direct control over the mesh size. This allows to keep the cell number almost constant throughout the simulation, thus avoiding the creation of extremely small elements that would impact on the allowable timestep needed to keep the maximum Courant number below 1.

As the system under investigation is a planetary gearbox, the kinematics involves not only the pure rotation of the planet carrier, but also the roto-translation of the planets. This means that the GRA$^{MC}$ cannot be applied directly. However, after the first engagement between the planets and the ring gear has occurred, it is possible to reuse the first mesh by applying to it a rigid rotation of an angle $\theta$ given by:

$$\theta = \frac{2\pi}{z_{ring}} \tag{7}$$

By exploiting this artifice, the enormous advantages of the GRA$^{MC}$ on the computational time can still be exploited, and the set of mesh that covers one complete planet–ring gear engagement can be used for the entire simulation. From the mesh deformation tests, it emerged that a set of 10 grids is sufficient to complete one engagement. Therefore, every 10 meshes a rigid rotation of multiples of $\theta$ was applied to the initial mesh set, thus providing the necessary meshes for the whole simulation. The rigid rotation of the mesh is a simple operation that requires less than one second on a modern hardware, while the remeshing of the computational domain for every position would be much more computationally expensive. In Figure 4 the workflow of the adopted mesh-handling strategy is illustrated.

As it can be noticed, the procedure foresees the mesh-to-mesh interpolation of the results. With each successive domain being conformal, the mapping is perfectly consistent. Furthermore, the mesh set is composed of very similar grids in terms of element size and count, hence the interpolation errors are minimized. The whole simulation process is automated in a Bash [34] script. The base case can be adapted to investigate the desired operating conditions by changing the velocity and the time libraries consistently.

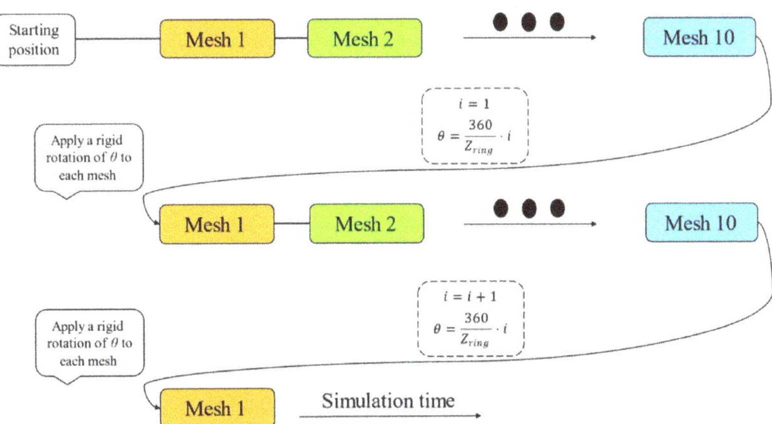



**Figure 4.** Flowchart of the implemented mesh-handling strategy.

### 2.4.4. Boundary Conditions and Numerical Settings

The main kinematic characteristic of a planetary gearbox and is that the planets perform a roto-translation around the center axis. This means that the velocity of the planets is zero at the pitch point with the ring gear, and it increases linearly in the radial direction. The velocity contour plot that describes the kinematics is shown in Figure 5.

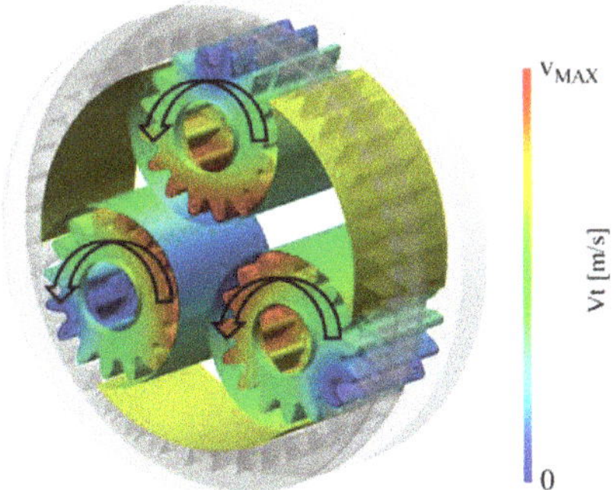



**Figure 5.** Velocity contour plot of the system in the central position.

A dedicated boundary condition that applies the roto-translation was implemented and assigned to the three planets. The planet carrier is assigned a boundary condition that applies a pure rotation around its center axis. The housing and the ring gears are fixed in space and do not undergo any deformation. With it being a closed system, the pressure and the volumetric fraction fields have a Neumann boundary condition, i.e., must be calculated. Wall adhesion models can be included by specifying the contact angle parameter (angle formed by a liquid at the three-phase boundary where liquid, gas, and solid intersect). Factors such as porosities, surface roughness, and impurities can affect this parameter influencing the wettability of the wall surfaces. In the current work, the contact angle was not considered, and the Neumann boundary condition (*zeroGradient*) was applied to the alpha field on the walls.

The PIMPLE (merged PISO-SIMPLE) algorithm was used since it exploits the possibility to use relatively high time steps with relaxation (SIMPLE) and to maintain the temporal information (PISO). The time step was set to 0.1° rotation of the planet carrier. In Table 5, the numerical settings used in the simulation are summarized.

**Table 5.** Numerical settings used in the simulations.

| | |
|---|---|
| Convergence criterion | $1 \times 10^{-5}$ |
| Maximum Courant number | 1 |
| Pressure solver | PCG (preconditioned conjugate gradient) |
| Velocity solver | PBiCG (stabilized preconditioned bi-conjugate gradient) |
| Time derivative discretization | First order implicit Euler scheme |
| Velocity discretization | Second order linear-upwind scheme |
| Convection of the volumetric fraction | Second order van Leer scheme |

### 2.4.5. Computational Performance

A 48 GFLOPs Linux Workstation (INTEL Xeon® Gold 6154 CPU, 4 Cores, 3 GHz) was used to run the simulations, which required on average 20 h each for a single planet carrier rotation. The low computation time achieved by parallelizing the computations on only 4 cores can be attributed to the newly implemented remeshing procedure, which is based on the recursive use of the initial mesh. This way, the effort of the remeshing process is drastically reduced.

Without the mesh-handling strategy explained in Section 2.4.3 (computation of a limited mesh set and application of rigid rotations), the computational domain would have needed to be partitioned and meshed continuously for every gear position, with a significant increase in the computational effort. In Table 6, the implemented GRA$^{MC}$ with a multi-rotation strategy is compared with a general approach that does not consider the repeatability of the tooth positions.

**Table 6.** Comparison of the computational effort between a standard and the newly implemented remeshing algorithm.

| | GRA$^{MC}$ | General Approach | Net Gain % |
|---|---|---|---|
| Simulation time for the analyzed gearbox (1 planet carrier rotation) | 20 h | 400 h | 95% |

By the developed procedure, a 95% time gain per planet carrier rotation could be achieved. This is mainly because the grids were previously computed and are immediately available for the mapping process as the simulation continues. Moreover, owing to the proper domain partitioning that was carried out in the pre-processing phase, it was possible to mesh the geometry with a series of extrusions starting from a 2D mesh. The resulting 3D elements are triangular prisms, which exhibit higher quality parameters and require less memory consumption with respect to tetrahedrons, thus resulting in faster calculation times. Liu et al. [2] simulated the same planetary gearbox with a local remeshing approach. Their simulation model, which also considered the roller bearings and a high level of detail, implemented in ANSYS Fluent, required 15–20 h parallelized on 80 cores of a high-performance computing cluster.

## 3. Results and Discussion

In this chapter, the results of the simulations and the high-speed camera recordings of the experimental investigations for the selected operating points are shown. Firstly, a direct comparison with the data coming from high-speed camera recordings is presented. Secondly, the views taken from different perspectives were extracted from the CFD models to describe the oil flow inside the planetary gearbox.

### 3.1. Comparison of Numerical and Experimental Results

Exemplary simulation results with respect to the oil flow were compared with high-speed camera recordings to validate the oil distribution in the front part of the gearbox. The simulation results are shown after one rotation of the planet carrier, as no appreciable changes in the oil flow were observed moving on with the simulation. Figures 6–8 show a side-by-side comparison for the low filling level operating conditions. Dashed colored rectangles are used to highlight and emphasize similarities between the experimental and numerical results.

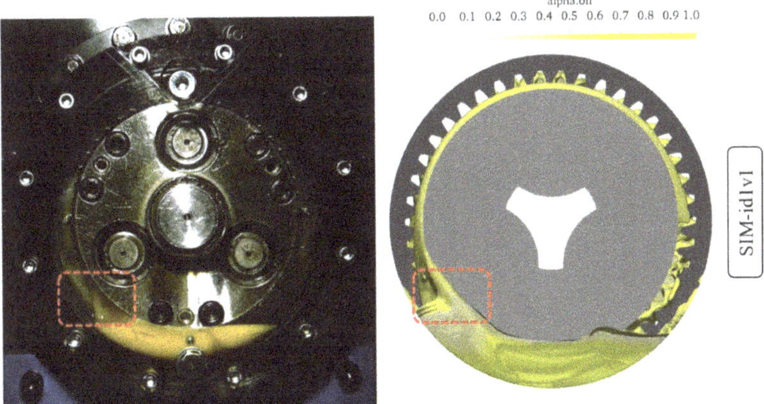

**Figure 6.** Experimental (**left**) and CFD (**right**) comparison of the oil flow for SIM-id1v1.

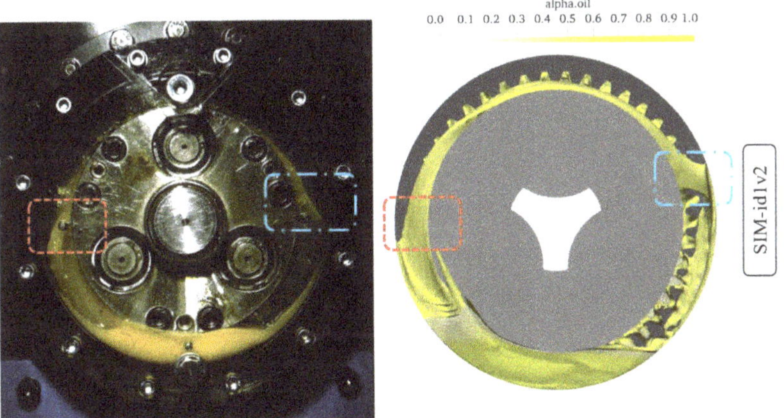

**Figure 7.** Experimental (**left**) and CFD (**right**) comparison of the oil flow for SIM-id1v2.

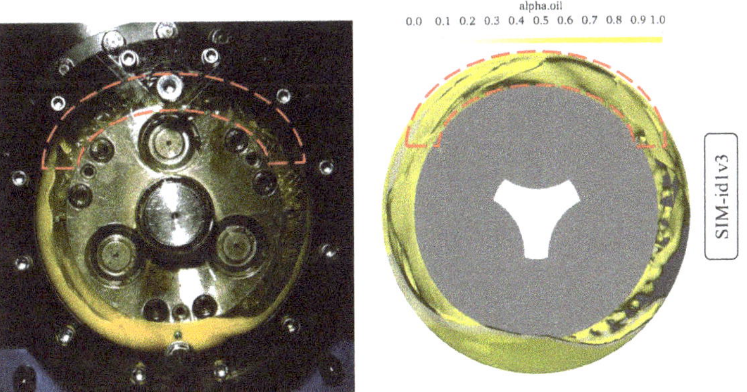

**Figure 8.** Experimental (**left**) and CFD (**right**) comparison of the oil flow for SIM-id1v3.

For the SIM-id1v1 operating condition (Figure 6), the low rotational speed translates into a low amount of oil being dragged by the planet carrier rotation. This results in the formation of an oil track in the lower left region (red dashed rectangle) and only a small quantity is carried up by the planet carrier rotation. This behavior is well captured by the simulation. As the velocity increases (SIM-id1v2, Figure 7), more oil is dragged by the planet carrier. This reflects into higher splashing on the left area (red dashed rectangle), as well as into a pronounced oil ejection on the right side of the gearbox (blue dashed dot rectangle). Both phenomena are very well predicted by the CFD model. At the highest velocity (SIM-id1v3, Figure 8), the centrifugal forces promote a marked radial expansion of the oil towards the housing, also including the region above the planet carrier (red dashed arc). However, the simulation predicts a greater amount of oil in the upper region. A possible explanation for this difference could be related to the modeling of wall adhesion at the contact point between the wall surface and the interface, as explained in Section 2.4.4. These three operating conditions were investigated by Liu et al. [2] as well. Moreover, in their model, the planet carrier rotation drags the oil from the sump and brings it to the upper part of the system. It was observed that the higher the rotational speed, the higher the amount of oil being splashed around. The fact that their model includes the bearings can introduce some differences in the small-scale flows in that region. Dedicated sub-models with a much finer mesh around the bearings and specific boundary conditions for the correct description of the rings' and rollers' kinematics would be necessary to accurately predict the oil flow and the velocity and pressure gradients in the bearings region [35].

Figures 9–11 show a side-by-side comparison for the high filling level operating conditions.

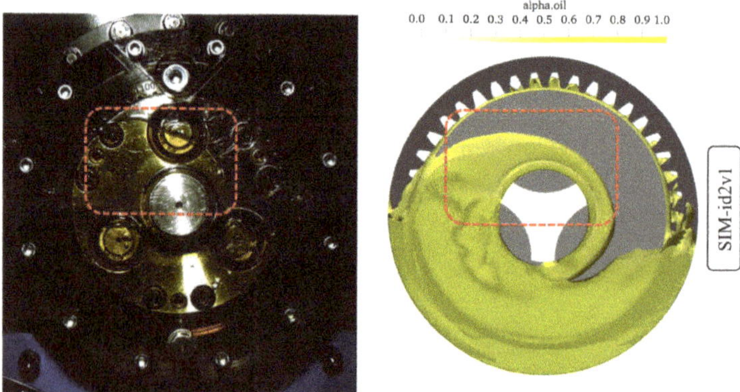


**Figure 9.** Experimental (**left**) and CFD (**right**) comparison of the oil flow for SIM-id2v1.

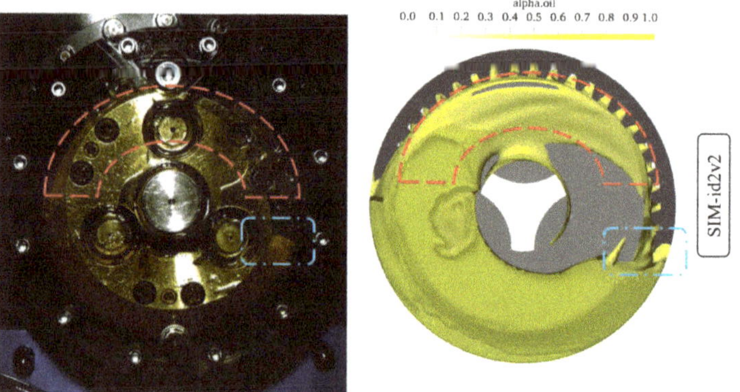


**Figure 10.** Experimental (**left**) and CFD (**right**) comparison of the oil flow for SIM-id2v2.

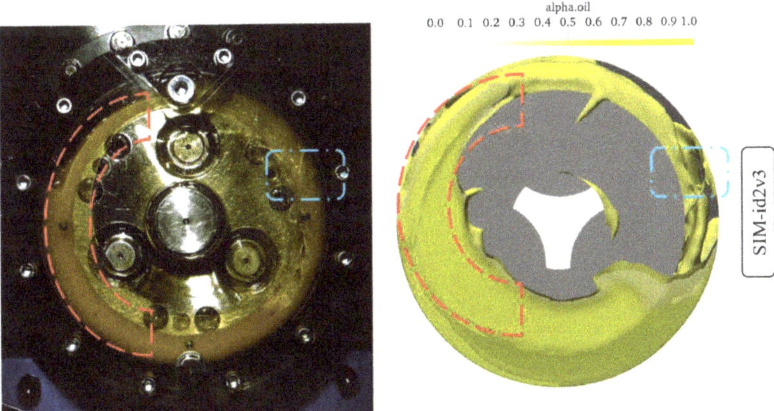


**Figure 11.** Experimental (**left**) and CFD (**right**) comparison of the oil flow for SIM-id2v3.

For the SIM-id2v1 operating condition (Figure 9), from the high-speed camera recordings it can be observed that a peculiar shape of the oil originates: these are Taylor–Couette flows [36] that arise between the rotating planet carrier and the fixed internal surfaces of the housing, as demonstrated by De Gevigney et al. [7]. The wave that can be seen in front of the planet carrier is well predicted by the CFD model (red dashed rectangle). As the velocity increases (SIM-id2v2, Figure 10), the wave shape of the oil tends to cover a larger area in the gearbox (red dashed arc). A slight decrease in the oil sump level could be observed on the right side of the planet carrier (blue dashed dot rectangle). At the highest velocity (SIM-id2v3, Figure 11), similarly to SIMI-id1-v3, the centrifugal forces create an evident oil ring around the planet carrier (red dashed arc), in accordance with the observations by Boni et al. [8]. The oil is splashed all around the gearbox (blue dashed dot rectangle), resulting in a completely different behavior with respect to the other investigated velocities, in which the oil is also present in the front region. The absence of the oil in front of the planet carrier and its splashing around the gearbox is well predicted by the model. However, differently from the experimental recordings, in the CFD model of SIM-id2v1 and SIM-id2v2 more oil is sticking to the planet carrier and less oil is falling into the sump. A possible explanation for this difference could be related to the modeling of wall adhesion at the contact point between the wall surface and the interface, as explained in Section 2.4.4. For SIM-id2v3, a larger amount of oil in the red dashed arc is predicted by the simulation if compared with the experimental recordings, but the shape and the flow path of the oil have been captured.

In this section, the numerical results were compared with the experiments. The transparent housing allowed to observe the flow that originates in the front part of the gearbox. In this regard, the oil rings that originate from the planet carrier rotation are already evident from the shown side-by-side comparison. Additional oil rings originate inside the gearbox around the planets, as will be presented from the CFD results in the following paragraph.

*3.2. Details of the Oil Flow in the Planetary Gearbox*

To gain deeper insight into the oil flow inside the gearbox, 3D isometric views are reported for the analyzed operating conditions. The gearbox housing was made transparent to enable access to the internal fluid domain region. In Figure 12, the oil distribution is shown for the analyzed operating conditions. In the top and in the bottom row of the figure, the low and the high filling level cases are shown, respectively. Near each result, the corresponding operating condition is indicated in a box on the bottom right.

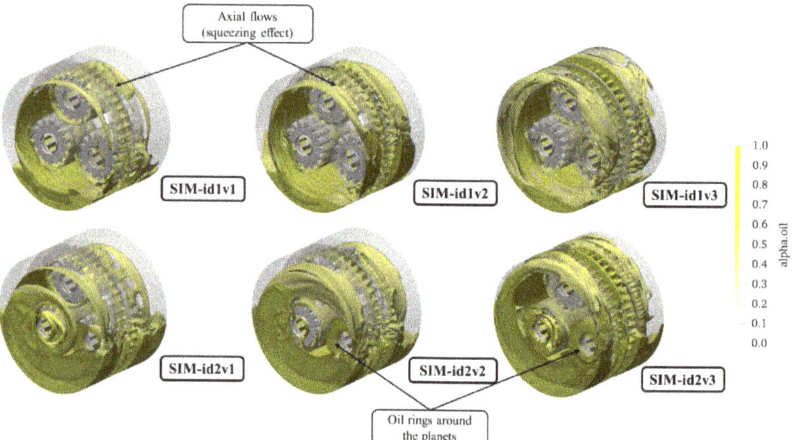



**Figure 12.** Oil distribution inside the gearbox for the analyzed operating conditions. Top row: low filling level cases. Bottom row: high filling level cases. Axial flows (squeezing effect) and oil rings around the planets are highlighted.

For the low filling level cases (top row), the oil is dragged from the sump and delivered to the other parts of the gearbox. As the rotational velocity increases, more oil participates to the lubrication and the ring gears receive a greater amount of oil. For the high filling levels (bottom row), oil rings also originate around the planet gears, as illustrated in Figure 12. These are generated by the roto-translation of the planets which are completely submerged when they are below the centerline of the gearbox. As they move above the centerline, the oil remains entrapped between the planets and the planet carrier, and the roto-translation of the wheels promotes the formation of these additional oil rings. The faces of the gears remain wet, and the cylindrical shape of the gear shafts contribute to the rotatory movement of the oil around the wheels.

The squeezing effect that can be seen in all cases derives from the mutual interaction between the planets and the ring gears. The squeezing of the oil from the planet–ring gear meshing zone was also observed by Boni et al. [8]. The CFD simulations provide a physical explanation for these axial fluxes. Indeed, while the gears engage, the gap between the teeth reduces and increases continuously during the gear mesh. The sudden contraction of this volume implies an overpressure in the gap that gives origin to axial fluxes. After reaching the minimum value for the volume, the volume of the gap suddenly increases again causing a lower pressure in the gap (it should be considered that the pressure drop is significantly smaller than environment pressure of 1 bar, and, therefore, this pressure gradient will not lead to phase exchanges, namely, cavitation). It should be highlighted that the calculation of the pressure gradients due to the volume variation during the gear meshing with the present model is not representative of the pressures on the mating flanks near the contact area. These considerations are confirmed in literature [37–41]. In Figure 13, the pressure contour in the mating region that justifies the oil squeezing is reported. The 2D representation shows the slice taken in the center of the middle gear.

In Figure 14, the evolution of the oil flow at different timesteps during the planet carrier rotation is illustrated for SIM-id2v3. From the standstill position (A), the planets gradually enter in contact with the oil in the sump (B, C, D). The oil is then distributed circumferentially along the ring gears (E, F). At this velocity, the planet carrier rotation discards the oil radially towards the housing. The oil that remains in the sump is recovered by the wheel's passage during the motion and is continuously brought to the other parts of the gearbox.

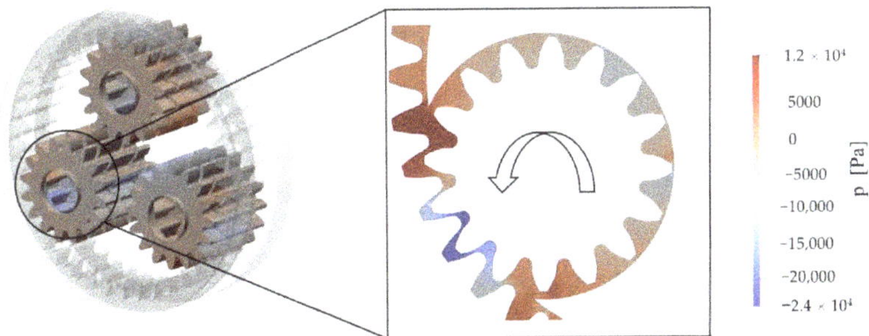

**Figure 13.** Pressure contour on the wheels.

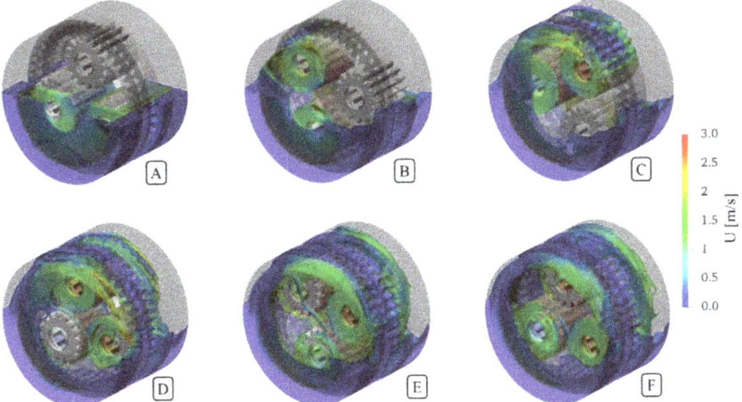

**Figure 14.** (**A–F**) evolution of the oil flow for the SIM-id2v3 operating condition at different timesteps during the planet carrier rotation. The oil is gradually brought to the other parts of the gearbox. The contact between the planets and the ring gears seems to be well lubricated. The rotatory motion of the oil flow gives origin to oil rings around the planet carrier and the planets.

In this section, a better insight into the lubrication in the planetary gearbox was provided. The oil squeezed between the planets and the ring gears creates axial flows that meet the oil coming from the planet carrier rotation, confirming what was observed by Boni et al. [0]. Additional oil rings around the planets were detected. The reason behind these fluxes could be explained by the layout and kinematics of planetary gearboxes, in which rotation and revolution motions coexist.

## 4. Conclusions

In the present work, a planetary gearbox was analyzed with a combined experimental–numerical approach. A transparent housing cover allowed recording of the oil flow in the front part of the gearbox by a high-speed camera. This allowed to capture the oil flow characteristics that typically originate in planetary gearboxes. The current analysis proposes a new remeshing approach that is on the one hand based on the proper partitioning of the computational domain to obtain an extruded mesh and, on the other hand, based on an innovative mesh-handling strategy that exploits a pre-computed set of mesh for the whole simulation. The implemented approach allowed running of such simulations on only four cores without the need for a computer cluster. The simulation results are in good agreement with the experimental observations, despite some discrepancies at the centerline immersion depth, where the oil remained stitched on the planet carrier in the simulations and did not flow down. The simulations at lower immersion depth could be compared with Liu et al. [2]. The oil flow prediction in the front part of the gearbox is similar to the one predicted in their work. However, a different level of detail can introduce some differences in the oil flow.

The current work allowed a good understanding of the oil flow characteristics in the considered planetary gearbox: at low speeds, oil rings originate because the oil remains in contact with the rotating parts and this phenomenon promotes circular oil tracks; at high speeds, the oil is mainly splashed towards the housing due to the high centrifugal effects and forms a ring around the housing walls. These observations refer to the investigated immersion depths. These phenomena have also been observed by Liu et al. [2] in their analysis at different immersion depths and oil viscosities.

The numerical studies on planetary gears available in the literature are mostly not validated so far. In this study, for the first time, experimental observations and CFD modeling were successfully combined to evaluate the numerical approach used to predict oil distribution and contribute to a better understanding of oil flow in planetary gears.

**Author Contributions:** M.N.M.: conceptualization, methodology, software, validation, formal analysis, investigation, resources, data curation, writing—original draft, writing—review and editing, visualization. L.H.: validation, investigation, writing—review and editing. C.P.: conceptualization, validation, investigation, writing—review and editing. T.L.: conceptualization, writing—review and editing, supervision, project administration. K.S.: writing—review and editing, resources. F.C.: conceptualization, methodology, software, validation, formal analysis, investigation, resources, data curation, writing—original draft, writing—review and editing, visualization, supervision. All authors have read and agreed to the published version of the manuscript.

**Funding:** The publication of this work is supported by the Open Access Publishing Fund of the Free University of Bozen/Bolzano.

**Institutional Review Board Statement:** Not applicable.

**Informed Consent Statement:** Not applicable.

**Data Availability Statement:** The data that support the findings of this study are available from the corresponding author (F.C.) upon reasonable request.

**Acknowledgments:** The authors would like to thank Paolo Codeluppi and Giuseppe Boni from Comer Industries for sharing the experimental data.

**Conflicts of Interest:** The authors declare no conflict of interest.

## References

1. Schudy, J. Untersuchungen Zur Flankentragfähigkeit von Außen-Und Innenverzahnungen. Einflüsse Auf Das Grübchen-, Grauflecken- Und Verschleißverhalten, Insbesondere Bei Langsam Laufenden Getriebestufen. Ph.D. Thesis, Technical University of Munich, Munich, Germany, 2010.
2. Liu, H.; Standl, P.; Sedlmair, M.; Lohner, T.; Stahl, K. Efficient CFD Simulation Model for a Planetary Gearbox. *Forsch. Ing.* **2018**, *82*, 319–330. [CrossRef]
3. Gold, P.W.; Hermsmeier, J.; Goedecke, O.; Assman, C. Untersuchung Der Leerlaufverluste in Einem Planetengetriebe Für Windkraftanlagen Mit Mine-Ralölbasischen Und Biologisch Schnell Abbaubaren Schmierstoffen. *VDI Berichte* **1999**, *1460*, 217–230.
4. Kettler, J. *Ölsumpftemperatur von Planetengetrieben*; FVA Heft 639; FVA: Frankfurt am Main, Germany, 2002.
5. da Costa, D. Power Loss in Planetary Gear Transmissions Lubricated with Axle Oils. Master's Thesis, University of Porto, Porto, Portugal, 2015.
6. Höhn, B.R.; Stahl, K.; Schudy, J.; Tobie, T.; Zornek, B. FZG Rig-Based Testing of Flank Load-Carrying Capacity Internal Gears. *Gear Technol.* **2012**, 60–69.
7. de Gevigney, J.D.; Changenet, C.; Ville, F.; Velex, P.; Becquerelle, S. Experimental Investigation on No-Load Dependent Power Losses in a Planetary Gear Set. In Proceedings of the International Gear Conference, Valencia, Spain, 4–5 March 2013; pp. 1101–1112.
8. Boni, J.-B.; Neurouth, A.; Changenet, C.; Ville, F. Experimental investigations on churning power losses generated in a planetary gear set. *J. Adv. Mech. Des. Syst. Manuf.* **2017**, *11*, JAMDSM0079. [CrossRef]
9. Boni, J.-B.; Changenet, C.; Ville, F. Analysis of Flow Regimes and Associated Sources of Dissipation in Splash Lubricated Planetary Gear Sets. *J. Tribol.* **2021**, *143*, 111805. [CrossRef]

10. Maccioni, L.; Concli, F. Computational Fluid Dynamics Applied to Lubricated Mechanical Components: Review of the Approaches to Simulate Gears, Bearings, and Pumps. *Appl. Sci.* **2020**, *10*, 8810. [CrossRef]
11. Bianchini, C.; Da Soghe, R.; Errico, J.D.; Tarchi, L. Computational Analysis of Windage Losses in an Epicyclic Gear Train. In Proceedings of the Turbo Expo, Charlotte, NC, USA, 26–30 June 2017; American Society of Mechanical Engineers (ASME): New York, NY, USA, 2017; Volume 5B.
12. Bianchini, C.; Da Soghe, R.; Giannini, L.; Fondelli, T.; Massini, D.; Facchini, B.; D'Errico, J. Load Independent Losses of an Aeroengine Epicyclic Power Gear Train: Numerical Investigation. In Proceedings of the ASME Turbo Expo 2019: Turbomachinery Technical Conference and Exposition, Phoenix, AZ, USA, 17–21 June 2019.
13. Concli, F. Austempered Ductile Iron (ADI) for gears: Contact and bending fatigue behavior. *Procedia Struct. Integr.* **2018**, *8*, 14–23. [CrossRef]
14. Concli, F.; Conrado, E.; Gorla, C. Analysis of power losses in an industrial planetary speed reducer: Measurements and computational fluid dynamics calculations. *Proc. Inst. Mech. Eng. Part J J. Eng. Tribol.* **2013**, *228*, 11–21. [CrossRef]
15. Concli, F. Thermal and efficiency characterization of a low-backlash planetary gearbox: An integrated numerical-analytical prediction model and its experimental validation. *Proc. Inst. Mech. Eng. Part J J. Eng. Tribol.* **2015**, *230*, 996–1005. [CrossRef]
16. Concli, F. Low-Loss Gears Precision Planetary Gearboxes: Reduction of the Load Dependent Power Losses and Efficiency Estimation through a Hybrid Analytical-Numerical Optimization Tool [Hochleistungs- Und Präzisions-Planetengetriebe: Effizienzschätzung Und Reduzierun. *Forsch. Ing.* **2017**, *81*, 395–407. [CrossRef]
17. Concli, F.; Gorla, C. Numerical modeling of the churning power losses in planetary gearboxes: An innovative partitioning-based meshing methodology for the application of a computational effort reduction strategy to complex gearbox configurations. *Lubr. Sci.* **2017**, *29*, 455–474. [CrossRef]
18. Concli, F.; Gorla, C. CFD Simulation of Power Losses and Lubricant Flows in Gearboxes. In Proceedings of the American Gear Manufacturers Association Fall Technical Meeting 2017, Colombus, OH, USA, 22–24 October 2017; pp. 2–14.
19. OpenFOAM. Available online: https://www.openfoam.com (accessed on 20 December 2022).
20. Cho, J.; Hur, N.; Choi, J.; Yoon, J. Numerical Simulation of Oil and Air Two-Phase Flow in a Planetary Gear System Using the Overset Mesh Technique. In Proceedings of the Open Archives of the 16th International Symposium on Transport Phenomena and Dynamics of Rotating Machinery, ISROMAC 2016, Honolulu, HI, USA, 10–15 April 2016.
21. Benek, J.; Steger, J.; Dougherty, F.C. A Flexible Grid Embedding Technique with Application to the Euler Equations. In Proceedings of the 6th Computational Fluid Dynamics Conference, Danvers, CO, USA, 13–15 July 1983; p. 1944.
22. Benek, J.; Buning, J.; Steger, J. A 3-D Chimera Grid Embedding Technique. In Proceedings of the 7th Computational Physics Conference, Cincinnati, OH, USA, 15–17 July 1985; p. 1523.
23. STAR-CCM+. Available online: www.plm.automation.siemens.com (accessed on 8 December 2022).
24. ANSYS FLUENT. Available online: www.ansys.com (accessed on 8 December 2022).
25. Höhn, B.R.; Stahl, K.; Schudy, J.; Tobie, T.; Zornek, B. Investigations on the Flank Load Carrying Capacity in the Newly Developed FZG Back-to-Back Test Rig for Internal Gears. In Proceedings of the Fall Technical Meeting; AGMA: Cincinnati, OH, USA, 2011.
26. Laukotka, E.M. *Referenzöle—Datensammlung*; FVA Heft 660; FVA: Frankfurt am Main, Germany, 2003.
27. Versteeg, H.K. *An Introduction to Computational Fluid Dynamics—The Finite Volume Method*; Pearson Education: London, UK, 1995.
28. Hirt, C.W.; Nichols, B.D. Volume of fluid (VOF) method for the dynamics of free boundaries. *J. Comput. Phys.* **1981**, *39*, 201–225. [CrossRef]
29. Rusche, H. *Computational Fluid Dynamics of Dispersed Two-Phase Flows at High Phase Fractions*; Imperial College of Science, Technology and Medicine: London, UK, 2002.
30. Farrell, P.E.; Maddison, J.R. Conservative interpolation between volume meshes by local Galerkin projection. *Comput. Methods Appl. Mech. Eng.* **2011**, *200*, 89–100. [CrossRef]
31. Python. Available online: https://www.python.org/ (accessed on 8 December 2022).
32. SALOME. Available online: http://www.salome-platform.org (accessed on 8 December 2022).
33. Mastrone, M.N.; Concli, F. CFD simulations of gearboxes: Implementation of a mesh clustering algorithm for efficient simulations of complex system's architectures. *Int. J. Mech. Mater. Eng.* **2021**, *16*, 1–19. [CrossRef]
34. Bash. Available online: www.gnu.org/software/bash (accessed on 8 December 2022).
35. Concli, F.; Schaefer, C.T.; Bohnert, C. Innovative Meshing Strategies for Bearing Lubrication Simulations. *Lubricants* **2020**, *8*, 46. [CrossRef]
36. Taylor, G.I. VIII. Stability of a viscous liquid contained between two rotating cylinders. *Philos. Trans. R. Soc. London Ser. A* **1923**, *223*, 289–343. [CrossRef]
37. Concli, F.; Della Torre, A.; Gorla, C.; Montenegro, G. A New Integrated Approach for the Prediction of the Load Independent Power Losses of Gears: Development of a Mesh-Handling Algorithm to Reduce the CFD Simulation Time. *Adv. Tribol.* **2016**, *2016*, 1–8. [CrossRef]
38. Concli, F.; Gorla, C. A CFD analysis of the oil squeezing power losses of a gear pair. *Int. J. Comput. Methods Exp. Meas.* **2014**, *2*, 157–167. [CrossRef]
39. Korsukova, E.; Morvan, H. Preliminary CFD Simulations of Lubrication and Heat Transfer in a Gearbox. In Proceedings of the ASME Turbo Expo 2017: Turbomachinery Technical Conference and Exposition GT2017, Charlotte, NC, USA, 26–30 June 2017.

40. Fondelli, T.; Massini, D.; Andreini, A.; Facchini, B. Three-Dimensional CFD Analysis of Meshing Losses in a Spur Gear Pair. In Proceedings of the ASME Turbo Expo 2018: Turbomachinery Technical Conference and Exposition GT2018, Oslo, Norway, 11–15 June 2018.
41. Li, J.; Qian, X.; Liu, C. Comparative study of different moving mesh strategies for investigating oil flow inside a gearbox. *Int. J. Numer. Methods Heat Fluid Flow* **2022**. [CrossRef]

*applied sciences*

*Article*

# A Comparison between Two Statistical Methods for Gear Tooth Root Bending Strength Estimation Starting from Pulsator Data

Luca Bonaiti [1,2,*], Michael Geitner [2], Thomas Tobie [2], Carlo Gorla [1] and Karsten Stahl [2]

[1] Politecnico di Milano, Dipartimento di Meccanica, Via La Masa 1, 20156 Milano, Italy
[2] Gear Research Center (FZG), Technical University of Munich, 85748 Garching bei München, Germany
* Correspondence: luca.bonaiti@polimi.it

**Abstract:** Due to their cost-effectiveness, pulsator tests are widely adopted as a testing methodology for the investigation of the effects of material and heat and surface treatment on the gear strength with respect to tooth root fatigue fracture. However, since no meshing contact is present in pulsator tests, there are differences between the test case and the real-world application scenario where gears are rotating under load. Those differences are related to both statistical and fatigue phenomena. Over the years, several methodologies have been developed in order to handle this problem. This article summarizes them and proposes a first comparison. However, no complete comparison between the different estimation methodologies has been conducted so far. This article aims to partially cover this gap, first by presenting and comparing the methodologies proposed in the literature and then via a deeper comparison between two different elaboration methodologies. Those two methodologies, which have been developed by examined to the same test rig configuration, are also discussed in detail. The comparison is performed based on an actual database composed of 1643 data points from case-hardened gears, divided into 76 experimental campaigns. Good agreement between the estimated gear strengths was found. The database is also adopted in order to make further considerations about one methodology, providing additional validation and defining the specimen numerosity required.

**Keywords:** gears; fatigue in gears; gear testing; tooth root bending fatigue; SN curve

**Citation:** Bonaiti, L.; Geitner, M.; Tobie, T.; Gorla, C.; Stahl, K. A Comparison between Two Statistical Methods for Gear Tooth Root Bending Strength Estimation Starting from Pulsator Data. *Appl. Sci.* **2023**, *13*, 1546. https://doi.org/10.3390/app13031546

Academic Editor: Qingbo He

Received: 9 December 2022
Revised: 17 January 2023
Accepted: 20 January 2023
Published: 25 January 2023



## 1. Introduction

Gears are machine elements that, as a result of the meshing of their profiles, allow the transmission of power between two rotary axes while maintaining a (nominal) constant ratio between the rotational velocity of the two axes [1–3]. Their working principle implies the presence of a loaded sliding/rolling contact condition that is located far from the base of the tooth (i.e., the tooth root). The contact implies the co-presence of several failure modes, such as tribological damages (i.e., scuffing and wear) and fatigue damages (e.g., (macro)pitting, micropitting, tooth flank fracture, and tooth root fatigue fracture). The latter (also called tooth root bending fatigue or tooth root breakage) is related to the pulsating normal force that acts on the tooth flanks, which causes a bending stress history in the tooth root that starts from a small negative value (when the previous tooth pair is in mesh) up to a maximum (ideally when the contact is at the outer point of the single tooth contact) [4,5]. Furthermore, the root radius works as a notch, further increasing the maximum stress [5]. Amongst all the failures that can affect a gear, tooth root fatigue fracture is considered to be the most critical one. Indeed, the failure of a single tooth implies the stoppage of the power flow within the gearbox and typically leads to a total failure of the gear system.

Fortunately, standards such as ISO 6336-3 [6] and ANSI AGMA 2001-D04 [7] provide gear designers with standardized analytical tools that can be used to assess a gear pair with respect to this failure mode. The assessment is based on the comparison of the occurring tooth root stress with the permissible tooth root stress. The first is primarily based on the maximum tensile stress at the surface in the tooth root fillet, typically for external gears at the 30° tangent, and depends on the gear pair meshing characteristics and the tooth root shape. The latter is based on typical gear material (together with the treatments) strength data and is defined at 1% gear failure probability. Typical resistance data are also part of these standards (i.e., [7,8]).

However, due to their intrinsic limitations, the standards cannot cover all possible varieties of materials and (heat or surface) treatments. Moreover, as Figure 1 shows, even the standards do not provide a comprehensive gear SN curve but rather a scattering range (see the shaded area) for $>3*10^6$ load cycles, leaving the designer to assume strength in the long-life region. Indeed, the standards themself strongly recommend performing experimental campaigns to properly estimate the gear load-carrying capacity.

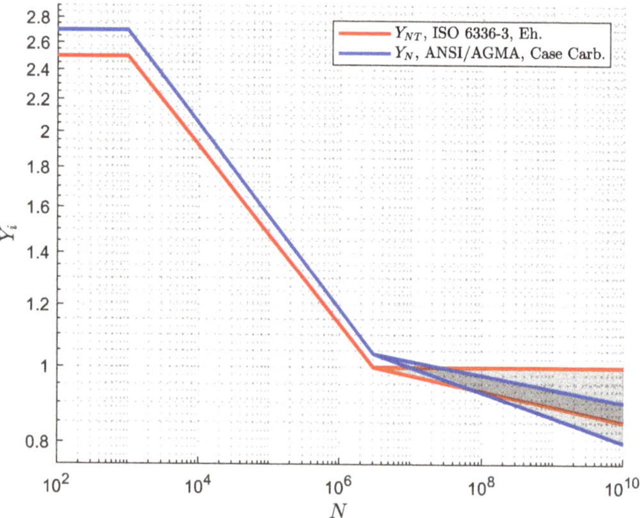


**Figure 1.** Typical shape of gear SN curve according to ISO 6336-3 [6] and ANSI AGMA 2001-D04 [7] considering tooth root bending for case-hardened gears.

Three main testing methodologies for tooth root fatigue fracture are present in the literature: The Meshing Gear (MG) test (e.g., [9]), the pulsator test (e.g., [10–15]), also known as the Single Tooth/Teeth Bending Fatigue (STBF) test, and the test of notched specimens (e.g., [16–20]). The most common testing methodology is the pulsator test [21]; indeed, the tests are performed on a gear, rather than a notched specimen, while using a generic uniaxial testing machine, rather than a full gearbox. Therefore, pulsator testing is a cost-effective method that considers a representative test specimen. A description of the typical testing rig is presented in Section 2.

ISO 6336-5 [8] strength number data have been obtained based on both industrial experience and experimental investigations applying this kind of testing methodology. In this context, it is important to mention that although pulsator tests are part of the ISO 6336 series, the standard does not provide any indications about how to perform pulsator experimental campaigns and how to elaborate their outcome.

The fact that in pulsator tests, the teeth are loaded by anvils (clamping jaws) rather than by gear meshing, implies that there are some differences between the pulsator test and the real-world application scenario. Those differences are related to two main aspects:

1. From the fatigue point of view, the tooth root stress history is different. On the one hand, in pulsator tests, the stress trend is sinusoidal with R > 0. On the other hand, in the MG case, the tooth root stress presents a peculiar stress trend that is influenced by the load sharing between teeth pairs (and the related tooth deformability). This

difference is typically solved by means of corrective coefficients that modify the SN curve (e.g., [22]).

2. From the statistical point of view, it must be considered that pulsator tests are performed on selected teeth rather than on the complete gears. Furthermore, not all teeth of a gear can be tested. Commonly, statistical tools are adopted here.

Pulsator tests are also affected by the typical difficulties of every fatigue experimental campaign, that is, the proper estimation of the experimental SN curve, both in terms of test planning and how to determine the SN curve while also taking into account the runouts (i.e., specimens that overcome the runout level, that is, the number of cycles after which the test is suspended).

To overcome all these difficulties, several methodologies have been proposed in the literature. However, no systematic and holistic comparison between the outcomes of different estimation methodologies has been made so far. This paper aims to partially cover this gap.

First, a comparison between methodologies is proposed here. Then, focusing only on the statistical aspect of the pulsator problem, a more detailed comparison between the two estimation methodologies is discussed. This is performed by examining the gear SN curves estimated starting from the pulsator test data. Those two methodologies were developed by examining a symmetric test configuration together with a specific emphasis on case-hardened gears. One method is currently adopted at FZG (Gear Research Center, in German: *Forschungsstelle für Zahnräder und Getriebesysteme*) and is described in [23–26]. The second one, instead, had been recently developed at *Politecnico di Milano* (POLIMI from now on) [27,28]. Both gear SN curve estimation methods have been developed by the authors.

Here, after a short presentation of the testing methodology (Section 2) and a literature review of how the literature deals with pulsator test data (Section 3), two estimation procedures are described deeply in Sections 4 and 5. The comparison between the two methodologies was performed based on the case-hardened gears experimental database presented in Section 6. The comparison is reported in Section 7. In Section 8, the adopted database is used to further evaluate the general behavior of the second model.

In this paper, all the mathematical discussions are discussed in the appendix for the sake of simplicity. Furthermore, as the problem of a different tooth root stress history is not discussed here, the term "gear SN curve" refers to the gear under pulsator test loading conditions (and not under MG conditions).

## 2. Experimental Procedure for Tooth Root Bending Fatigue Testing

Pulsator or STBF tests are performed with the idea of loading one or two teeth at a fixed location to ensure that tooth root bending fatigue is the only present failure mechanism. This approach leads to a variety of testing configurations such as the one adopted by Seabrook and Dudley [29], where torque is applied to a toothed shaft that is in contact with a fixed anvil, the three-point bending configuration (e.g., [1,30]), and the one-tooth (also known as asymmetric) and two-teeth pulsator (also known as symmetric) test configurations; the last two seem to be the most commonly adopted.

The one-tooth test configuration was first described by Buenneke et al. in 1982 [31] and is now included in the SAE recommendation practice J1619 [32]. Here, the force is applied only on a single tooth while another tooth and a centering pin provide the reaction to the applied load. Therefore, only a single tooth is fully loaded. Hence, this configuration is typically called asymmetric (the same is applied here). It is worth mentioning that STBF (Single-Tooth Bending Fatigue) has been developed to describe this type of test. However, now this term is also adopted as synonymous with a pulsator test (e.g., [15,33]), therefore describing all the possible test methodologies that aim to perform tooth root bending fatigue tests by loading the gear teeth (or tooth) by an anvil(s) (rather than in MG conditions).

In the two-teeth configuration, the load is transmitted from one tooth to the other without the interaction of any other parts of the testing equipment. Here, the force is applied by two clamping jaws, typically spanning over three to five teeth. As a result of the principle involved in the Wildhaber measurement [34] and by proper design of the text fixture (i.e., the distance between the anvils and the gear hub centerline equal to the span measure over two), these jaws load the teeth at the same time at the same nominal diameter in a direction that is tangential to the base diameter. Therefore, both the tested teeth are subject to the same nominal tooth root bending stress. Hence, this configuration is typically called symmetric (the same is applied here). This configuration can still be referred to as STBF but, differently from the previous case, represents Single-Teeth Bending Fatigue. Research is now underway with the aim to investigate the effect of dynamic loads on tooth root stress via a specific testing rig (e.g., [35]).

Here, it is worth mentioning that by properly setting the distance between the anvils and the gear hub centerline, it is possible to obtain an asymmetric two-teeth test configuration (e.g., [33]). This configuration is adopted to maximize the toot root stress of one of the two tested teeth; it is rarely used.

Figures 2 and 3 show two examples of pulsator machines working in symmetrical configurations. All the data discussed here have been obtained with this methodology.

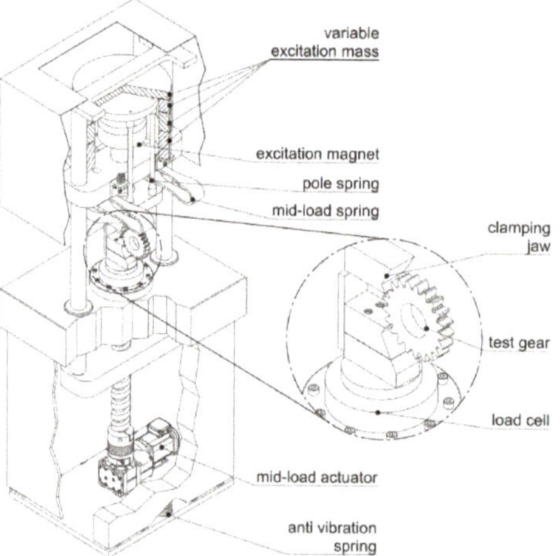


**Figure 2.** Pulsator test rig (schematic) adopted at FZG [36].

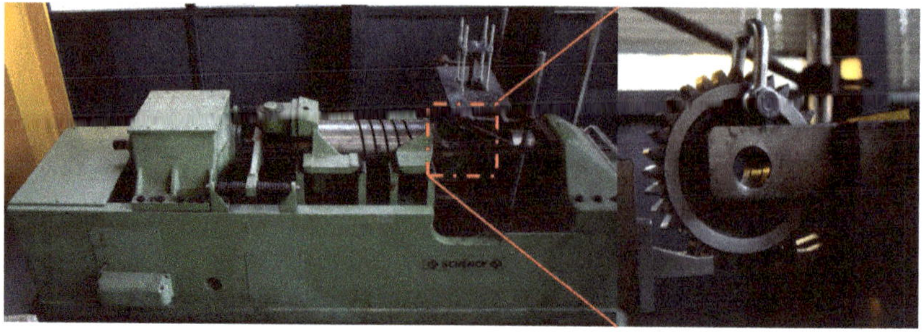

**Figure 3.** The Schenck mechanical resonance pulsator adopted at POLIMI.

To maintain the specimen in position, pulsator tests must be run with a fixed preload. For instance, the SAE recommendation practice J1619 [32] suggests performing tests at

$R = 0.1$, while within FVA guideline no. 563 I [25] a load ratio between 0.03 and 0.075 is suggested.

### 3. The Pulsator to Running Gears Problem

In the previous section, the typical experimental pulsator test rig configurations were presented. The fact that gear teeth are loaded by jaws rather than by rotating gear meshing implies that differences occur between the pulsator test case and the actual application case of MG. Recently, Hong et al. [37], whose testing rig is an asymmetric one, summarized those differences as follows:

1.  Pulsator tests and MG have different loading conditions, as, in the first, a sinusoidal load is applied at a fixed position, while in the latter, the load changes both the magnitude (due to the load sharing between teeth pairs) and the point of application (as the contact point between teeth changes during gear meshing).

2.  The stress ratio R is different as pulsator tests are required to work at $R > 0$, while in the MG gear case, the tooth root stress moves to a positive maximum (typically when the contact is at the outer point of a single tooth in contact) to a negative minimum (due to the extension of the compression field from the adjacent tooth, e.g., [4,5,38]), thus at a load ratio $R < 0$ (e.g., [39]).

3.  The crack initiation location seems to be different. Indeed, certain experimental findings of Winkler et al. [40] and Fuchs et al. [36,41–44] possibly imply that in the pulsator test case, crack initiation is much more likely to occur at the surface, while in the MG case, a slightly increased risk for subsurface crack initiation can be observed. However, their findings are based on investigations on shot-peened gears made from high-cleanliness steels. Furthermore, a different runout level was considered for pulsator and MG tests. Moreover, considering all the data available [36,40–44], and further results from experimental investigations at FZG that have not been published yet, it can be assumed that there is not a different crack location for common case-hardened gear steels for both compared testing methods (i.e., pulsator and MG). Therefore, the effect of the crack initiation location will not be discussed here.

4.  The statistical behavior of pulsator and MG tests is different. In pulsator tests, the failed teeth are predetermined and are those that are loaded within the pulsator test rig (for each test run, two teeth are in the symmetric configuration); furthermore, not all the gear teeth can be tested. On the other hand, in the MG case, all the gear teeth are loaded during a test run, and the gear failure is the result of the failure of the weakest gear tooth (e.g., [22,29,37,45–47]).

Therefore, strength numbers determined by pulsator tests should be elaborated to ensure a reliable estimation for gear design. In 1964, with a slightly different test rig, Seabrook and Dudley [29] already suggested that pulsator experimental strength numbers should be reduced by a factor of 0.77 to account for those differences.

In 1987, Rettig [22] dealt with the problem of the different loading conditions and stress ratio between the test methods. He suggested that, in order to consider the differences in the loading conditions, pulsator test results should be reduced by a factor of 0.9. The validity of Rettig's coefficient has also been reaffirmed recently by Concli et al. [48–50], who adopted high-cycle multiaxial fatigue criteria to estimate the effect of the different stress histories occurring at the tooth root.

In 1999, Rettig's work [22] laid the groundwork for the estimation procedure reported by Stahl et al. (FVA research project 304) [23]; this approach is discussed in detail in Section 4. This method is completed by the FVA guideline 563 I [25], in which additional information (e.g., the sampling strategy) is provided to the experimenter. According to Stahl et al. [23], pulsator tests should be reduced by a factor of 0.83 for peened and 0.77 for unpeened gears. Those values are completely comparable to what Seabrook and Dudley [29] found.

In 2003 and 2008, Rao and Mc. Pherson [45,46] proposed a different approach. On the one hand, they used allowable stress range diagrams to evaluate the effect of the different load ratios. On the other hand, statistical correlations, similar to the concepts of the return period, were used in order to deal with the statistical problem. A modified staircase procedure was adopted to estimate the experimental procedure. They also observed a difference between the pulsator and MG resembling the one discussed by Seabrook and Dudley [29]. By extension, Rao and Mc. Pherson's [45,46] findings are also similar to what Stahl et al. [23] calculated.

Between 2021 and 2022, another framework was proposed in [27,28,51]. Two different tools were adopted. Firstly, due to a different load ratio than the one adopted by Rettig [22], high cycle multiaxial fatigue criteria were adopted to deal with the different stress trends [51]. Then, in [27,28], the maximum likelihood method (ML) was adopted to estimate the experimental SN curve. Finally, by means of Statistics of Extremes (SoE), the gear SN curve was estimated. This approach is discussed in detail in Section 5. As will be shown in the following sections, this method yields results similar to that of Stahl et al. [23].

In 2022, Hong et al. [37] also proposed a different estimation procedure. They developed corrective coefficients in order to deal with the first three differences (i.e., the loading condition, stress ratio, and crack initiation location) while statistical relationships (similar to Rao and Mc. Pherson [45,46]) were exploited to develop a fourth coefficient dealing with the statistical differences. They also adopted ML to estimate an experimental curve based on the model of Pascual and Mekker [52]. However, as also discussed by Hong et al. [37], this methodology still needs to be further improved.

It is worth also mentioning two recent works that focus on the topic of experimental pulsator data elaboration. The first is the one of Mao et al. [53], where they developed corrective coefficients to overcome some limitations of the traditional Dixon–Mood method (whose description can be found in [54]). Similarly, to reduce the required specimen numerosity, Alnahlaui and Tenberge [55] proposed a small-sample-size test program based on the linear damage accumulation framework. However, in both works, no further discussions had been made about the estimation of the gear strength starting from pulsator data.

Additional information about the different methodologies and their differences can also be found in [27,37].

## 4. Method 1: FVA Approach Based on FZG Research

The first estimation method discussed here had been developed at FZG and is defined as a recommended procedure within FVA (Research Association for Drive Technology, in German: *Forschungsvereinigung Antriebstechnik e.V.*) research according to FVA guideline 563 I [25]. In [25], a practical approach for the evaluation of the tooth root load-carrying capacity, including the gear specimen, pulsator equipment, and test planning, was described. It is based mainly on the work of Stahl et al. (FVA research project 304) [23], where data elaboration is discussed, laying the fundamentals on the analysis of a large number of experimental data performed by FZG. The methodology according to [25] has been developed by examining a symmetric pulsator test configuration for case-hardened gears. Further investigations and detailed statistical analysis of an extended experimental FZG database can be found in Hein et al. [24,26] and Geitner et al. (FVA research project 610 IV) [56]. Refs. [24,56] also reaffirmed the validity of this methodology.

The FVA approach based on FZG research aims to estimate a gear SN curve such as the one described in Figure 4. Here, different considerations are made regarding the limited-life and long-life regions. On the one hand, for each load level, the first region presents a different lognormal deviation of the failure load cycle number. In other words, the dependency of the variance with respect to the load level is considered (see also Figure 6). On the other hand, the long-life region is described by an endurance limit, whose dispersion is in the load/stress direction. For both the experimental and gear SN curves, these two regions are estimated separately using two different calculation procedures.

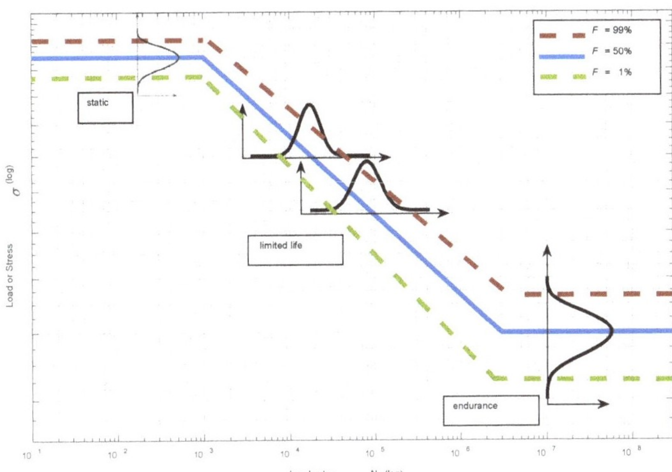



**Figure 4.** SN curves for different failure probabilities by assumption of separated consideration of long-life (endurance) and limited-life ranges [24]. Extracted from AGMA 18FTM26, Reliability of Gears—Determination of Statistically Validated Material Strength Numbers, with the permission of the publisher, the AmericanGear Manufacturers Association, 1001 North Fairfax Street, Suite 500, Alexandria, Virginia 22314.

According to this method, the experimental endurance limit is estimated according to Hück [57]. It is assumed that the experimental endurance limit follows a normal distribution. In the case of reduced sample size ($n < 10$), this method suggests estimating the experimental endurance limit by adopting the modified probit method as defined by Hösel and Joachim [58]. Figure 5 shows an example of the application of the Hück [57] approach, displaying a staircase test sequence together with an additional theoretical test run. The experimental endurance limit $\sigma_{F0\infty,50\%}$ is defined starting from the determination of the number of tests $f_i$ on load level $i$, leading to the sums $F$ and $A$.

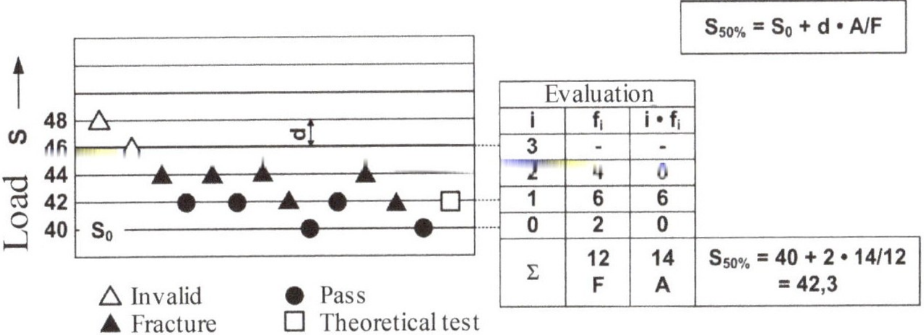



**Figure 5.** Exemplary evaluation according to Hück staircase method; example extracted from [25].

According to Hück [57], $\sigma_{F0\infty,50\%}$ is defined as:

$$\sigma_{F0\infty,50\%} = \sigma_{F0,i=0} + d \cdot \frac{\sum i \cdot f_i}{\sum f_i} = \sigma_{F0,i=0} + d \cdot \frac{A}{F} \tag{1}$$

Once the $\sigma_{F0\infty,50\%}$ has been estimated, the gear endurance limit $\sigma_{F0\infty,1\%,MG}$ (under the MG loading condition) is calculated by means of corrective coefficients [23]:

$$\sigma_{F0\infty,1\%,MG} = f_{1\%FD} \cdot f_{P \to MG} \cdot \sigma_{F0\infty,50\%,P} \tag{2}$$

where $f_{1\%FD}$ is equal to 0.92 for peened gears and 0.86 for unpeened gears according to Stahl [23]. An additional coefficient $f_{P\rightarrow MG}$ equal to 0.9, defined by Rettig [22], is adopted to take into account the load differences. $f_{P\rightarrow MG}$ was not applied in this paper; that is, $\sigma_{F0\infty,50\%}$ must be reduced by 0.77 or 0.83 to estimate $\sigma_{F0\infty,1\%,MG}$.

Within method 1, the limited-life region is estimated by calculating, for each load level, the 50% experimental lifetime $N_{50\%}$ by assuming a log-normal distribution of the failure load cycle numbers [23]. This is performed using a log-normal probability grid, with the definition of the single failure probability for each data point based on the order approach defined by Rossow [59]. On the other hand, a more simplified approach is to calculate $N_{50\%}$ from the lifetime of the n individual tests according to Equation (3) [25]:

$$\log(N_{50\%}) = \frac{1}{n}\sum_{i=1}^{n}\log(N_i) \tag{3}$$

That is, calculating the average lifetime.

Once the 50% curve has been estimated, the 1% gear failure probability lifetime $N_{1\%}$ is calculated for each load level as:

$$N_{1\%} = 10^{(\log(N_{50\%})-2.33\cdot s_{log}} \tag{4}$$

where $s_{log}$ is the typical logarithmic standard deviation, the trend of which is shown in Figure 6. Then, both the 50% experimental and the 1% gear curve are calculated by fitting a line through the estimated $\log(N_{50\%})$ and $\log(N_{1\%})$.

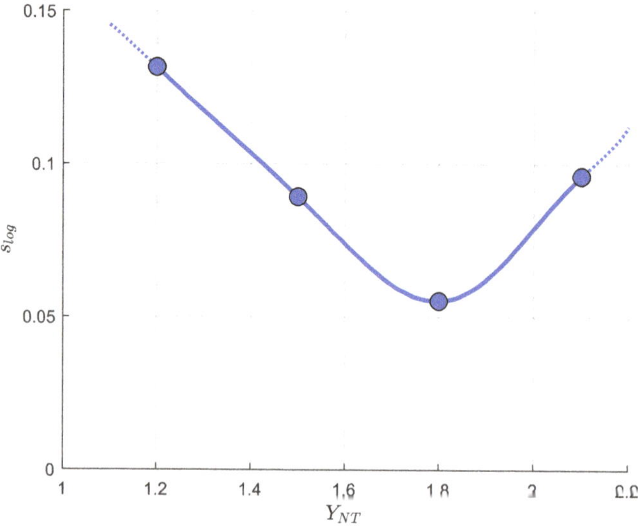


**Figure 6.** Logarithmic standard deviation considering failure load cycle numbers in limited-life range for peened gears according to Stahl [23].

According to this method, the standardized calculation method as per ISO 6336-3 [6] can still be applied. Therefore, the life factor $Y_{NT}$ is used to multiply $\sigma_{F0\infty,1\%,MG}$. The choice between a horizontal endurance line and the presence of the secondary slope for a number of load cycles > 3 million depends on the experience and the criticality within the field of application.

The FVA guideline 563 I [25] also describes the sampling strategy adopted (i.e., allocation and numerosity). According to this guideline, six different load levels are to be selected (if possible), two of which should be in the limited-life range. The load levels for the determination of the endurance limit must be equally distanced. Between 28 and

38 data points are suggested for a high numerosity experimental campaign. However, the standard allocation is between 20 to 24 test runs. The FVA guideline 563 I [25] also suggests that half of the specimens should be located in the endurance limit region while the remaining should be distributed among the load levels, aiming to have at least two load levels with a higher numerosity. Experimental campaigns with low numerosity are still possible, especially when the focus is primarily on the endurance limit. The runout level (i.c., the limiting load cycle number) is set to 6 million load cycles in pulsator tests.

An example of the application of this procedure is shown in Figure 7, where the dotted curve represents the 50% experimental SN curve while the continuous curve is the 1% gear curve. Here it is possible to observe the effect of the hypothesis of a not-constant deviation for the limited life. Indeed, the slopes associated with the limited life slightly vary. Similar to Figure 1, a shaded area has been included to represent the possible choices of $Y_{NT}$. The term "Exp" refers to the 50% curve estimated considering the experimental point while the term "Gear" refers to the gear SN curve (at 1% failure probability).

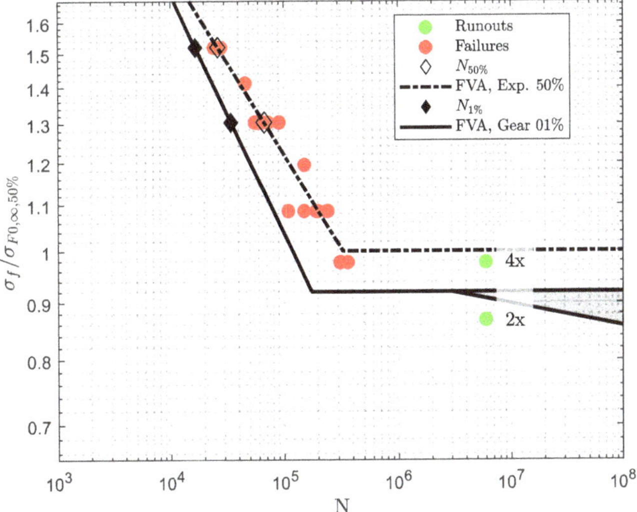


**Figure 7.** Evaluation of SN curves with different failure probabilities for exemplary campaign A from the experimental FZG database [24] according to FVA guideline 563 I [25].

## 5. Method I, Maximum Likelihood and Statistic of Extremes (ML&SoE)

Differently from the estimation approach presented in the previous section, in [27,28], a gear SN curve estimation approach was proposed based on two different tools: The ML method for the estimation of the experimental SN curve and SoE for its translation to the gear SN curve. Here, a two-slope curve formulation is adopted to describe the SN curve according to Spindel and Haibach [60]. According to this approach, the experimental SN curve can be described as:

$$\log(N) = \log(N_e) + \tfrac{1}{2}(k_1 + k_2)(\log(\sigma) - \log(\sigma_e)) + \\ + \tfrac{1}{2}(k_1 - k_2)|\log(\sigma) - \log(\sigma_e)| \tag{5}$$

where $N_e$ and $\sigma_e$ are the coordinate of the knee, and $k_1$ is the slope associated with the limited-life region and $k_2$ for the long-life region.

This model also features a constant standard deviation in the $\log(\sigma)$ direction $s_{\log(\sigma_{F0})}$ that, due to the first-order approximation [61], results in two different standard deviations in the $\log(N)$ direction $s_{1,\log(N)}$ and $s_{2,\log(N)}$ [60]. This model is also called the "bilinear uniform scatter band" (e.g., [62]). Data points are considered log-normally distributed, and differently from method 1 (see Section 4), all the points are considered part of a unique dataset. Therefore, here the distinction between limited-life and long-life regions depends only on the estimation outcome.

In order to estimate the teeth SN curve according to Equation (5), it is necessary to adopt the ML method due to the presence of survived specimens (i.e., specimens whose failure has not been observed) [61,63,64]. By considering the symmetric pulsator test configuration, depending on how the experimental data are considered, two different kinds of data can be obtained:

1. If the experimental points are considered as they are, that is, if the fact that the symmetric test configuration works on two teeth is not considered, the only observed surviving specimen is the one that reaches the runout level. The elaboration performed with such considerations is called "STBF" as they are the experimental points obtained directly from the pulsator/STBF testing machine.

2. If the fact that the symmetric test configuration works on two teeth is considered, it must be considered that the experimental point is not composed only of failed and surviving specimens but by the compresence of both. This can be easily understood by looking at the experimental points. On the one hand, if one of the two teeth fails (i.e., a failure according to the STBF consideration), the other has survived. That is, a failure occurring within the test is both an observed value (i.e., the failed tooth) and randomly right-censored data (i.e., the survived tooth). On the other hand, if the runout level is reached (i.e., a runout according to the STBF consideration), both tested teeth have reached the runout level. Therefore, different statistical considerations must be taken into account in the 2T case. The elaboration performed with such considerations will be called "2T" as two teeth are taken into consideration.

Details about the ML estimation of Equation (5) curve (together with $s_{\log(\sigma_{F0})}$) are reported in Appendix A. Once the initial curve parameters have been estimated, it is necessary to estimate the curve describing the gear. This is achieved by considering the hypothesis that in the case of tooth root bending fatigue, the gear fails when its weakest tooth breaks [22,29,37,45–47]. Therefore, SoE is used to elaborate the teeth Cumulative Distribution Function (CDF) to estimate the gear CDF. Through a mathematical passage, it allows for the estimation of the CDF describing the lowest value that can be observed by sampling n times from the same population. On the one hand, the term "lowest value that can be observed" is the weakest tooth (that describes the gear). On the other hand, the "same population" is represented by tested teeth and is described by a log-normal distribution whose mean is defined according to Equation (5) and standard deviation $\sigma_{\log(\sigma_{F0})}$.

When the STBF curve is studied, it is necessary to consider that this curve is based on a teeth pair, of which we observe the lifetime of the weakest one. By considering the gear as a system of $z/2$ teeth pair, it is possible to estimate the gear CDF $F_{gear}$ as:

$$F_{gear} = 1 - (1 - F_{STBF})^{z/2} \tag{6}$$

Similarly, when the 2T curve is studied, $F_{gear}$ is defined as:

$$F_{gear} = 1 - (1 - F_{2T})^{z} \tag{7}$$

Different curves at different reliability levels can be obtained by calculating the corresponding percentile for several load stages.

An example of the application of this procedure is shown in Figure 8; here, the experimental points are the same as in Figure 7. The dotted curve represents the curve that estimates the SN curve of the exemplary campaign according to the STBF and 2T approximation, while the continuous lines are the gear SN curves, estimated according to Equation (6) or Equation (7). On the one hand, the comparison between the STBF and 2T curves allows us to understand the role of the 2T consideration. The 2T curve is slightly above the STBF one. On the other hand, focusing on the gear SN curves, it is possible to notice that both estimation methods lead to (almost) completely identical curves.

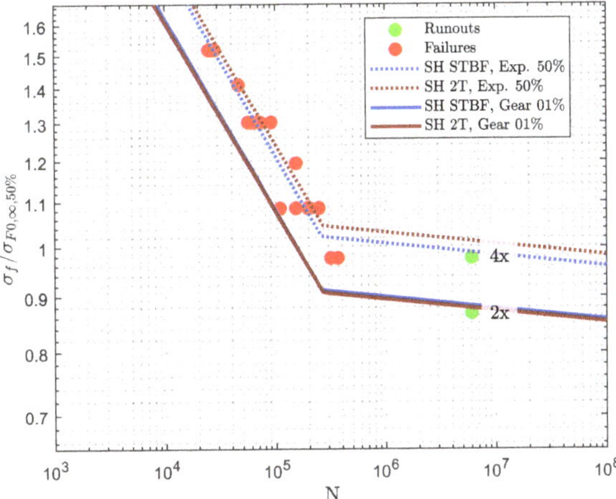


**Figure 8.** Evaluation of SN curves with different failure probabilities for exemplary campaign A from the experimental FZG database [24] according to the ML&SoE approach [27,28].

## 6. Adopted Database

The analysis presented in the following sections was carried out by taking advantage of the case-hardened gear database present at FZG, combined with the one presented by the gear research group at POLIMI. That is, all the data have been obtained by the authors' research groups.

FZG points have been partially described and analyzed by Hein et al. [24] considering the data contained in [40,65]. Focusing on the effects of materials and heat treatment properties, FZG data have been also analyzed in [56]. All associated FZG test campaigns described in [24] were performed after 2002 and were not available at the time of the work performed by Stahl et al. (FVA research project 304) [23]. More precisely, most of them have been performed in the last decade. POLIMI data points have been described in [27,28,66,67]. POLIMI test campaigns have been performed in the last two decades. Table 1 summarizes the database references in chronological order.

**Table 1.** Database reference summary.

| Year | Reference | Short Summary |
| --- | --- | --- |
| 2009 | [66] | Presentation and analysis of a part of POLIMI points |
| 2017 | [67] | Presentation and analysis of a part of POLIMI points |
| 2018 | [24] | Partial statistical analysis of FZG points |
| 2019 | [40] | Presentation and discussion of selected FZG campaigns |
| 2021 | [56] | Partial statistical analysis of FZG points with focus on effects of materials and heat treatments properties |
| 2022 | [65] | Presentation and discussion of selected FZG campaigns |
| 2021–2022 | [27,28] | Presentation and analysis of a part of POLIMI points |

Consequently, the analyzed test data represents up-to-date properties of commonly used case-hardened steels for gears. The database also includes tests supported by industrial sponsoring. Due to the confidentiality and comparability of the test campaigns, the exemplary test data are illustrated as anonymized and normalized. Furthermore, as will be subsequently mentioned in the paper, the test methodologies comparison has been performed on several further cases, which are not shown in this paper for the sake of simplicity. Figure 9 shows the database adopted here. It is composed of 1643 data points, divided into 76 experimental campaigns.

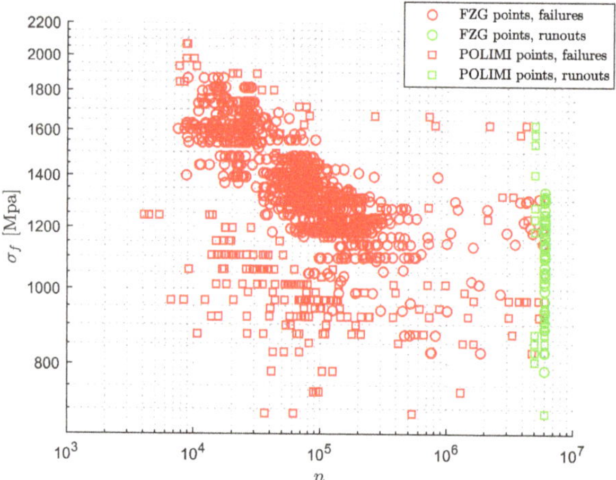

**Figure 9.** Pulsator test database [24,27,28,56,65–68].

## 7. Evaluation Study: Comparison between the Two Models

The simplest comparison between method 1 and method 2 is the direct comparison between the curves estimated with the two different approaches. Figures 10–14 show the outcomes of these two elaboration methods for five different pulsator test experimental campaigns. Several cases have been studied, obtaining similar results. For the sake of simplicity, only five of them are shown. For all the experimental campaigns, the solutions of method 2 have been verified by adopting the Likelihood Profile (LP) and Likelihood Ratio (LR). Experimental points shown in Figure 10 to Figure 13 were taken from the FZG database [24] while those in Figure 14 were obtained at POLIMI [27,28]. The SN curve comparison has not been extended to other Section 3 methodologies because they require specific data that are not available.

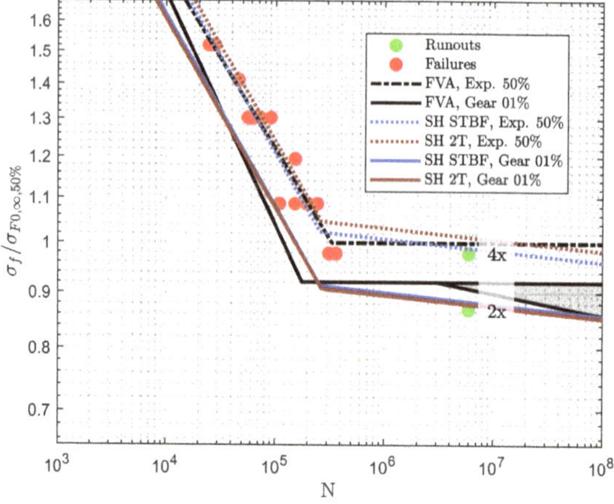

**Figure 10.** Comparison of two estimation methods for exemplary campaign A from experimental FZG database [24].

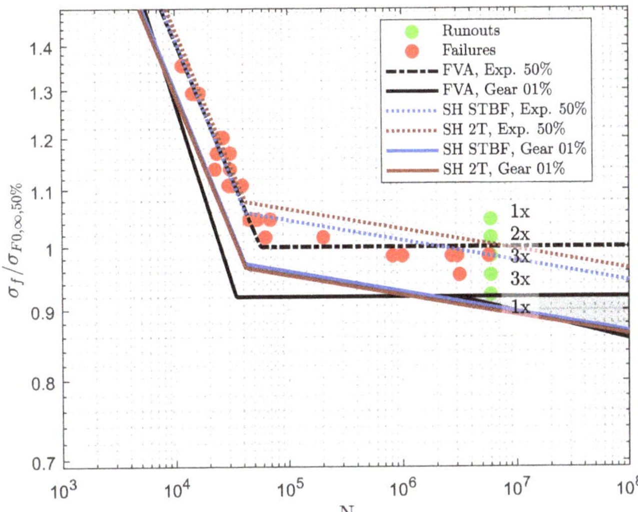

**Figure 11.** Comparison of the two estimation methods for exemplary campaign B from experimental FZG database [24,66].

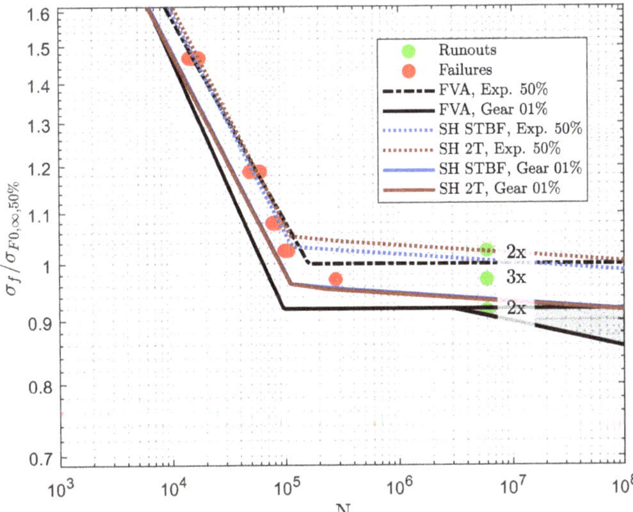

**Figure 12.** Comparison of the two estimation methods for exemplary campaign C from experimental FZG database [24].

By looking at the gear SN curves shown in Figures 10–14, it seems that there is no relevant difference between the two estimation techniques as the estimated curves are close to each other, especially if $Y_{NT}$ according to ISO 6336-3 [6] is selected on the conservative side in the long-life range. The good agreement between the models allows us to re-affirm the validity of the two models: Despite the usage of completely different statistical tools, the final gear curves are different but, most importantly, they are coherent with each other.

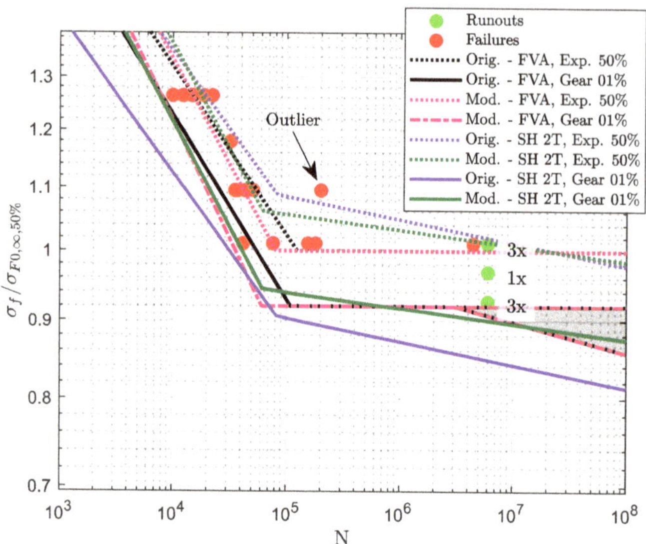

**Figure 13.** Comparison of the two estimation methods for exemplary campaign D from experimental FZG database [24] (Orig.) and excluding an outlier data point (Mod.).

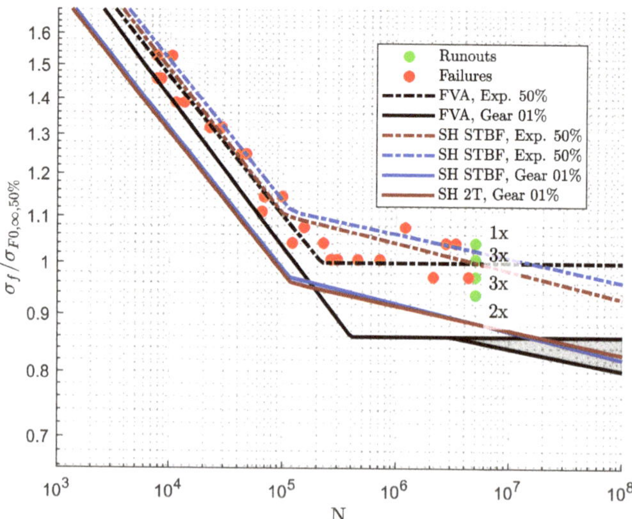

**Figure 14.** Comparison of the two estimation methods for exemplary campaign E from [27,28].

The first difference between the two estimation methods is the different positions of the fatigue knee. Especially for the Exp. curves, method 2 always estimates the knee above the one estimated by method 1. This different position is related to the different estimations of the complete SN curve. On the one hand, according to method 1, the fatigue knee is estimated as the intersection of the line describing the limited-life and the long-life regions. On the other hand, according to method 2, the fatigue knee location depends on the two estimated CDF parameters ($N_e$ and $\sigma_e$) and the CDF elaboration. However, neither method 1 nor method 2 present a smooth passage between the two regions. This aspect is taken into account within more complex SN curve estimation methods (e.g., [68–73]). Their application within the case of pulsator test elaboration is the subject of current research.

The main difference between the two estimation methods is the model sensitivity concerning the presence of outliers, and method 2 results in being more sensitive than method 1. This is depicted in Figure 13, where one experimental point (i.e., the one indicated by the arrow) can be considered an outlier as it has a lifetime almost 5 times higher than the one of the other experimental runs obtained at the same load level. This outlier has been defined by a graphical analysis of the test data and adopted here as an example.

On the one hand, concerning method 1, the only effect the outlier has is that, at such a load level, $N_{50\%}$ (i.e., Equation (3)) changes, with consequences on the estimated $N_{1\%}$ (i.e., Equation (4)); the lines describing the experimental and gear-limited life vary accordingly (see the relatively slight difference between the pink and the black line). As limited-life and long-life regions are estimated differently, this outlier does not affect the latter. Furthermore, as method 1 moves the estimated curves according to pre-defined coefficients (based on the typical scatter), the presence of an outlier has a minor effect as it affects only the estimated means. Indeed, both the original (i.e., the black one) and modified curves (i.e., the pink one) remain close to each other.

On the other hand, in method 2, it is possible to note how the identification of outliers is a crucial aspect. The two estimated curves are significantly different in both the limited-life and long-life regions, especially in the latter. Here, if the outlier, whose lifetime is the highest at the load level, is removed, the new gear SN curve indicates a higher resistance. This counterintuitive behavior is related to the estimation of $s_{\log(\sigma_{F0})}$. As a unique standard deviation is estimated by the model (i.e., $s_{\log(\sigma_{F0})}$), the presence of outliers implies an overestimation of $s_{\log(\sigma_{F0})}$. Such overestimation leads to a higher variance associated with the long-life region, with a detrimental effect on the gear SN curve. This aspect turns out to be the most crucial one. Furthermore, it is also possible to note how, as the model considers all the experimental points as a unique dataset, high load level points influence the long-life region (and vice versa).

Nevertheless, once the outlier has been removed, both methods 1 and method 2 give, once again, similar results.

## 8. Extended Investigations on Method 2

As mentioned in Section 4, the validity of method 1 had been reaffirmed in recent works (i.e., [24,56]). Both in [27] and here, the initial validation of method 2 was performed by comparing—partially—the outcome of the two models discussed here.

Further validation of method 2 can also be performed by comparing the outcome of the model parameters (together with their confidence interval) with typical gear data. Such validation is performed by examining the confidence intervals of $k_1$ and $k_2$. Confidence intervals are then compared with the typical values, which are the slopes of the limited-life and the long-life region for case-hardened materials according to ISO 6336-3 [6] and ANSI-AGMA 2001 [7].

The parameters $N_e$ and $\sigma_e$ were not considered. On the one hand, it is clear that in the database of Figure 9, the fatigue knee is not located at 3 million cycles (also see examples from Figures 10–14). On the other hand, discussion about $\sigma_e$ cannot be conducted for two reasons. Firstly, different from ISO $\sigma_{F\,lim}$ (i.e., the nominal bending stress number), $\sigma_e$ cannot be seen as either an endurance limit or as a fatigue limit. Indeed, $\sigma_e$ is just a parameter of Equation (5); therefore, it lacks all the statistical properties of an endurance (or fatigue) limit. Secondly, if one supposes that $\sigma_e$ is equivalent to $\sigma_{F\,lim}$ (or to ANSI-AGMA $s_{at}$), the comparison will become a comparison between the estimated material endurance limits and their standardized equivalent (which, in any case, is not the aim of this article).

The methodology adopted here takes inspiration from previous works. One example is the work of Beretta et al. [74] in which they performed a Monte-Carlo simulation of a fatigue database composed of 188 test data in order to investigate how to properly estimate a SN curve. Then, several testing conditions were simulated and the convergency of the parameter confidence interval was compared, aiming to define the best sampling strategy. A similar procedure was adopted by Loren and Lundtröm [75,76], where 100 staircases were virtually simulated to validate their models by comparing the estimates. Similar methodologies for the evaluation of the behavior of stress-lifetime models can also be found in [53,55,77,78].

Here, to increase the number of available experimental campaigns, as well as the numerosity of experimental campaigns with a high number of experimental points, some test series about similar gears made with comparable materials and treatments were combined. Twenty-seven new experimental campaigns were defined accordingly. Details about the confidence interval calculation according to LR are shown in Appendix B.

Figures 15 and 16 show a comparison between the estimated slopes (and their two-sided 95% confidence intervals) with their equivalent according to gear standards for both STBF and 2T. For most cases, the standardized values lie within the confidence interval. Based on this analysis, it is not possible to state that the estimated slopes are statistically different from the ones suggested by the standards. This behavior is particularly true for $k_1$ and $k_2$ according to ISO 6336-3 [6] and for $k_1$ and the highest (absolute) value $k_2$ according to ANSI-AGMA 2001 [7]. Therefore, it can be concluded that the parameters estimated by method 2 are not different from the typical ones of case-hardened gears.

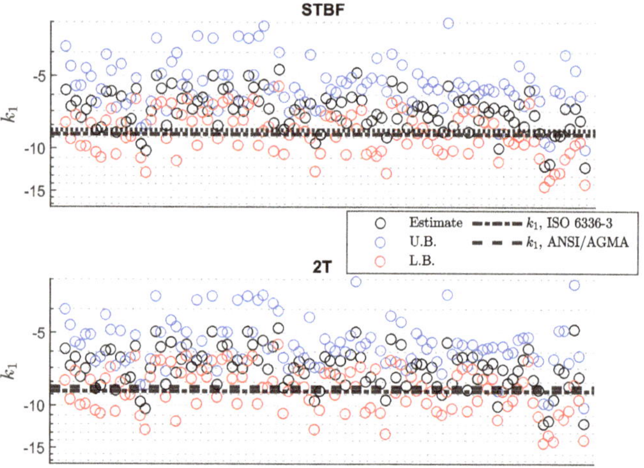

**Figure 15.** Comparison of estimated $k_1$ (and its confidence interval) with its standardized equivalent.

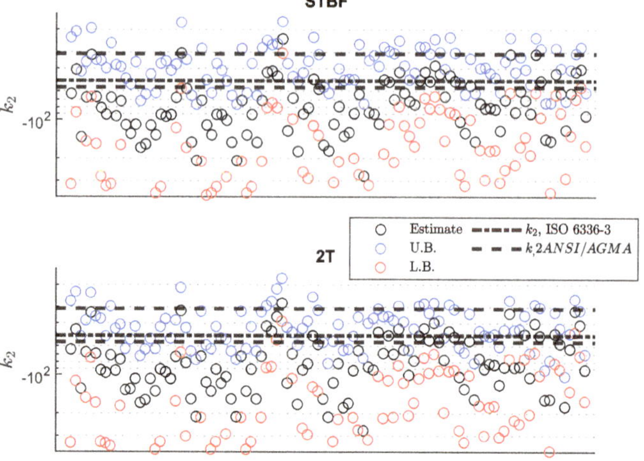

**Figure 16.** Comparison of estimated $k_2$ (and its confidence interval) with its standard equivalent.

As already mentioned, the FVA guideline 563 I [25] also provides indications about the allocation and numerosity of specimens. Those values have been defined within the context of data elaboration according to method 1. An indication of the number of experimental points required by model 2 can be obtained by taking inspiration from Beretta et al. [74].

Here, the Figure 9 database is used to evaluate the convergency of the parameter's confidence interval. Only the parameter $s_{log(\sigma_{F0})}$ is shown here. The rationale for this is that the variance is the crucial parameter for calculating the different percentiles. Furthermore, the other parameters all show an earlier convergency. As the database is composed of several materials, the comparison is performed by observing the normalized parameter.

Figure 17 shows the convergency of $\tilde{s}_{log(\sigma_{F0})}/\hat{s}_{log(\sigma_{F0})}$, that is, the convergency of the LR-based confidence interval $\tilde{s}_{log(\sigma_{F0})}$ (bilateral at 95%) normalized over the estimated one (i.e., $\hat{s}_{log(\sigma_{F0})}$). Both STBF and 2T cases are shown. In total, 103 experimental campaigns have been evaluated. Looking at the trend, after approximately 25–30 experimental points, there is no significant reduction in the confidence interval. Thus, 25–30 experimental points are the minimum number of points required for proper parameter estimation. In other words, 25–30 experimental points are required for a proper gear SN curve estimation if model 2 is adopted (30 to be on the conservative side).

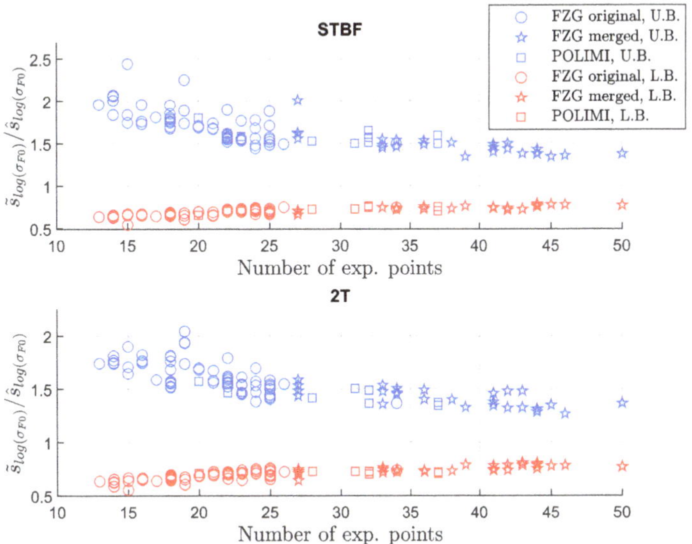


**Figure 17.** Trend of normalized $s_{log(\sigma_{F0})}$ confidence intervals.

## 9. Conclusions

Pulsator/STBF test is a test method employed in the estimation of gear tooth root load-carrying capacity with respect to the tooth root fatigue fracture phenomena. It is widely adopted because it is a cost-efficient method that considers a representative test specimen. Typical applications of pulsator tests are investigating the effects of material heat and surface treatments in gears. However, the pulsator test procedure presents several differences with respect to the real-world case where gears are meshing under load. Therefore, pulsator test results must be elaborated before using them to design a gear pair. Several methodologies have been developed over the years. Considering that all of them have been developed for different test configurations, good agreement has been found. That is, all of them suggest that, in the long-life region, pulsator test results should be reduced by a factor of approximately $\approx 0.77 - 0.83$.

Among all the methodologies, two of them have been developed by examining a symmetric (double teeth) pulsator test configuration. These two methodologies are discussed here in detail. The first (method 1: FVA, based on FZG research) elaborates pulsator test data based on typical scattering values obtained from a previous case-hardened gear test, that is, it adopts pre-determined coefficients. It must be noted that this methodology covers the whole testing procedure as additional information is also provided (e.g., gear specimen, testing machine). Most importantly, it also provides the experimenter with indications about the test planning. From a statistical and mathematical point of view, this methodology can be considered the simplest one. The latter (method 2: ML&SoE) adopts ML to estimate the experimental curve; then it estimates the gear SN curve by means of SoE. Different from the previous one, this methodology relies only on the observed data within the analyzed test series. However, no discussions about test planning have been performed so far for method 2, especially concerning specimen numerosity.

The two approaches discussed in this paper have been defined, independently and at different times, by all the authors of the paper. Both methodologies are described here in detail. In this way, the designer who wants to undertake the task of gear testing can find indications about the elaboration of pulsator test data.

The comparison of the methodologies has been performed on the basis of an actual database composed of 1643 symmetric pulsator test results obtained during 76 different experimental campaigns on case-hardened gears. The comparison was performed by directly examining the final gear SN curve. It has been found that both methods yield completely comparable results. The good agreement between the corresponding outcomes allows us to reaffirm the validity of the two models. However, as it does not rely on previously defined data, method 2 tends to be more sensitive to the presence of outliers. Several cases have been studied; for the sake of simplicity, only a few of them are reported in this paper.

As mentioned in Section 4, the validity of method 1 has been reaffirmed recently. Nevertheless, for method 2, both here and in previous works, the validation of method 2 has been performed only by comparing the outcomes of method 1 and method 2. Further validation of method 2 has been conducted here due to the availability of the case-hardened gear database. Parameter confidence intervals have been compared with typical gear data, finding good agreement between the model outcomes and their equivalents presented by the standards. Furthermore, the minimum number of points required for the application of the method has been calculated. This has been calculated as 30 to be on the conservative side.

**Author Contributions:** Conceptualization, L.B. and M.G.; methodology, L.B. and M.G.; software, L.B.; supervision, T.T., C.G. and K.S.; writing, L.B. and M.G. All authors have read and agreed to the published version of the manuscript.

**Funding:** This research received no direct external funding. However, the experimental database used includes results from several research projects over previous decades funded by different associations.

**Institutional Review Board Statement:** Not applicable.

**Informed Consent Statement:** Not applicable.

**Data Availability Statement:** Data are available normalized upon reasonable request.

**Conflicts of Interest:** The authors declare no conflict of interest.

**Acronym**

The following acronyms are used:

| | |
|---|---|
| CDF | Cumulative Distribution Function |
| ML | Maximum Likelihood |
| MG | Meshing Gear |
| LR | Likelihood Ratio |
| PF | Profile Likelihood |
| PDF | Probability Density Function |
| STBF | Single Tooth/Teeth Bending Fatigue |
| SoE | Statistics of Extremes |

**Nomenclature**

| | |
|---|---|
| $A$ | Parameter of Hück method |
| $d$ | Stress/load step, mainly considered in Hück method |
| $\gamma_1$ | Generic model parameter |
| $\hat{\gamma}_1$ | Estimated parameter |
| $\overline{\gamma_1}$ | Constrained parameter |
| $\tilde{\gamma}_1$ | Confidence interval of the estimated parameter |
| $F$ | Parameter of Hück method |
| $F_{gear}$ | Gear CDF |
| $F_{STBF}$ | STBF CDF |
| $F_{2T}$ | 2T CDF |
| $f_i$ | Number of tests on the $i^{th}$ load level, Hück's method |
| $f_{1\%FD}$ | Conversion factor acc. to Stahl for 50 % to 1 % failure probability |
| $f_{P \to MG}$ | Corrective coefficient acc. to Rettig |
| $k_1$ | Limited life slope |
| $k_2$ | Long life slope |
| $\mathcal{L}$ | Likelihood |
| $\ell$ | Log-likelihood |
| $N_{50\%}$ | 50% experimental lifetime |
| $N_{1\%}$ | 1% gear failure probability lifetime |
| $N_e$ | Spindel-Haibach model coordinate (together with $\sigma_e$) |
| $s_-$ | Standard deviation |
| $s_{log}$ | Logarithmic standard deviation |
| $s_{\log(\sigma_{F0})}$ | Spindel-Haibach standard deviation in the $\log(S)$ direction |
| $s_{1,\log(N)}$ | Spindel-Haibach standard deviation in the $\log(n)$ direction, limited life region |
| $s_{2,\log(N)}$ | Spindel-Haibach standard deviation in the $\log(n)$ direction, long life region |
| $\sigma_-$ | Stress |
| $\sigma_e$ | Spindel-Haibach model coordinate (together with $N_e$) |
| $\sigma_{F0\infty,50\%}$ | Experimental endurance limit for 50 % failure probability |
| $\sigma_{F0\infty,1\%,MG}$ | Gear endurance limit for 1 % failure probability |
| $Y_N$ | Life factor acc. to ANSI/AGMA 2001 |
| $Y_{NT}$ | Life factor acc. to ISO 6336-3 |

## Appendix A. Maximum Likelihood Estimation of the Experimental Tooth Root Fatigue Fracture SN Curve

ML is a parameter estimation technique based on finding distribution parameters that are most likely to describe the collected data. Mathematically, those estimated parameters are the ones that maximize the likelihood. Such an estimation technique is typically adopted to estimate the stress–life relationship of components (e.g., fatigue specimens, insulations, etc.). More complete details about this estimation method can be found in [61,63,64,79].

One of the greatest advantages of ML is its capability to deal with different kinds of data. Indeed, within the same statistical framework, it is possible to take into account, together with its statistical meaning, failures (i.e., observed data), survivals (i.e., right-censored data), failures occurring within different inspections intervals (i.e., interval-censored data), and failures occurring before the first inspection interval (left-censored data). In the case discussed here, only observed data and right censored data are present. For the Spindel–Haibach model (i.e., Equation (8)), the likelihood $\mathcal{L}$ can be described as:

$$\mathcal{L} = \prod_{i=1}^{n}\left(f_{x_i;\mu,\sigma}\right)^{\delta_i} \cdot \prod_{i=1}^{n}\left(1 - F_{x_i;\mu,\sigma}\right)^{1-\delta_i} \tag{8}$$

where $f_{x_i;\mu,\sigma}$ and $F_{x_i;\mu,\sigma}$ are the PDF/CDF describing the phenomena and $\mu$ and $\sigma$ are the parameters of the normal distribution (i.e., mean and standard deviation). $\delta_i$ is equal to 1 in the case of observed data and 0 in the case of right-censored data. $\prod_{i=1}^{n}\left(f_{x_i;\mu,\sigma}\right)^{\delta_i}$ is the term used to refer to observed data and $\prod_{i=1}^{n}\left(1 - F_{x_i;\mu,\sigma}\right)^{1-\delta_i}$ is the one referring to right-censored data. Here, $\mu$ and $\sigma$ are defined according to Section 5.

However, in order to facilitate the parameter estimation procedure, it is typical to work on the log-likelihood $\ell$ (i.e., the logarithm of $\mathcal{L}$), as the same properties apply:

$$\ell = \ln \mathcal{L} = \sum_{i=1}^{n}\delta_i(\ln f_{x_i}) + \sum_{i=1}^{n}(1 - \delta_i)(\ln(1 - F_{x_i})) \tag{9}$$

Considering a model with $p$ parameters $\gamma_1, \gamma_2, \ldots, \gamma_p$, the estimated parameters $\hat{\gamma}_1, \hat{\gamma}_2, \ldots, \hat{\gamma}_n$ are those that maximize the log-likelihood:

$$\hat{\ell} = \ell_{(\hat{\gamma}_1, \hat{\gamma}_2, \ldots, \hat{\gamma}_n)} = \max_{\gamma_1, \gamma_2, \ldots, \gamma_p} \ell_{(\gamma_1, \gamma_2, \ldots, \gamma_p)} \tag{10}$$

where $\hat{\ell}$ is the maximum log-likelihood.

From a practical point of view, as typical calculation software does not present maximization algorithms, $\hat{\gamma}_1, \hat{\gamma}_2, \ldots, \hat{\gamma}_p$ are estimated by minimizing $-\ell$.

## Appendix B. Profile Likelihood and Likelihood Ratio Profile

Profile Likelihood (PF) refers to the possibility of expressing how $\mathcal{L}$ (or $\ell$) depends on the parameters [61,63,64,80,81]. PF is formulated by constraining one (or more) parameter/s and calculating the new maximum likelihood. In other words, LR describes the trend of the constrained maximum likelihood in a certain range of the constrained parameter. However, as $\mathcal{L}$ cannot always be easily computed, it is typical to describe PF using $\ell$ rather than $\mathcal{L}$. If $\gamma_1$ is constrained to be equal to $\overline{\gamma}_1$, the constrained maximum log-likelihood for $\overline{\gamma}_1$ can be defined as:

$$\ell_{\overline{\gamma}_1} = \max_{\gamma_2, \ldots, \gamma_p} \ell_{(\gamma_1 = \overline{\gamma}_1, \gamma_2, \ldots, \gamma_p)} \tag{11}$$

An example of PF is shown in Figure A1, where the PF has been calculated for several values of $s_{\log(\sigma_{F0})}$.

The usage of PF allows for the description of the confidence interval of the parameter in its natural range [61,63,64] via the usage of the Likelihood Ratio (LR). It is possible to demonstrate that the ratio between $\hat{\ell}$ and $\ell_{\overline{\gamma}_1}$ can be related to a $\chi^2$ distribution with one degree of freedom [82]:

$$-2\ln R_{(\gamma_1)} = -2\ln\left(\frac{\mathcal{L}_{\overline{\gamma}_1}}{\hat{\mathcal{L}}}\right) = 2(\hat{\ell} - \ell_{\overline{\gamma}_1}) \sim \chi^2_{(1)} \tag{12}$$

According to Equation (12), it is possible to express the confidence interval of the parameter $\gamma_1$ with a $1 - \alpha$ confidence level (i.e., $\tilde{\gamma}_1$) by finding the value of $\overline{\gamma}_1$ that satisfies Equation (13) [61,63,64].

$$\ell_{\overline{\gamma}_1} = \hat{\ell} - \frac{1}{2}\chi^2_{(1-\alpha,1)} \tag{13}$$

Figure A1 also shows the confidence interval $\tilde{s}_{\log(\sigma_{F0})}$ evaluated according to Equation (13).

It is worth mentioning another utility of the LR framework. Indeed, most minimization tools such as (e.g., MatLab's *fminsearch*) require an estimation of the solution as an input parameter. By calculating the LP, several guess solutions are provided, thus allowing one to properly investigate, whereas the solver has calculated a proper solution.

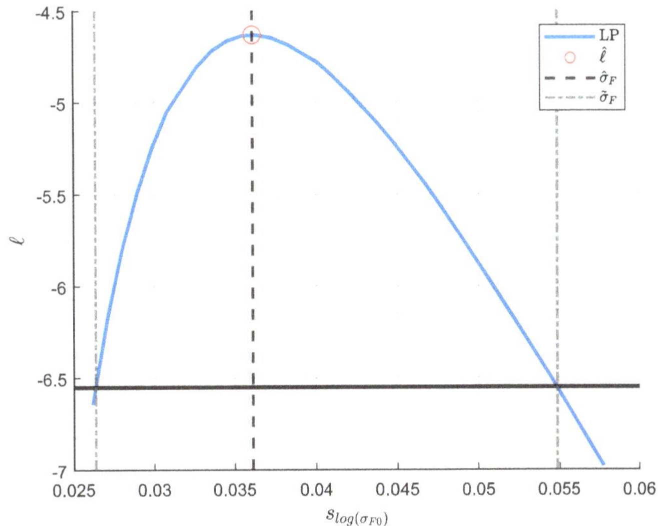

**Figure A1.** Example of LP and LR for $s_{log(\sigma_{F0})}$.

# References


1. Aktiengesellschaft, M. *Maag Gear-Book: Calculation and Manufacture of Gears and Gear Drives for Designers and Works Engineers*; Maag Gear-Wheel Co: Zurich, Switzerland, 1963.
2. Dudley, D.; Townsend, D. *Manuale Degli Ingranaggi - Edizione Italiana*; Tecniche nuove: Milano, Italy, 1996.
3. Henriot, G. *Manuale Pratico Degli Ingranaggi*; Tecniche nuove: Milano, Italy, 1993.
4. Fernandes, P. Tooth bending fatigue failures in gears. *Eng. Fail. Anal.* **1996**, *3*, 219–225. [CrossRef]
5. Davoli, P.; Conrado, E.; Michaelis, K. Recognizing gear failures. Mach. Desing. 2007. Available online: http://hdl.handle.net/11311/274596 (accessed on 14 January 2022).
6. *ISO 6336-3:2019*; Calculation of Load Capacity of Spur and Helical Gears—Part 3: Calculation of Tooth Bending Strength. International Organization for Standardization: Geneva, Switzerland, 2019. Available online: https://www.iso.org/standard/63822.html (accessed on 7 January 2022).
7. *ANSI AGMA 2001-D04*; Fundamental Rating Factors and Calculation Methods for Involute Spur and Helical Gear Teeth. American Gear Manufacturers Association: Alexandria, VA, USA, 2004. Available online: http://www.agma.org (accessed on 8 August 2022).
8. *ISO 6336-6:2019*; Calculation of Load Capacity of Spur and Helical Gears—Part 6: Calculation of Service Life under Variable Load. International Organization for Standardization: Geneva, Switzerland, 2019. Available online: https://www.iso.org/standard/63823.html (accessed on 26 January 2022).
9. Hong, I.J.; Kahraman, A.; Anderson, N. A rotating gear test methodology for evaluation of high-cycle tooth bending fatigue lives under fully reversed and fully released loading conditions. *Int. J. Fatigue* **2019**, *133*, 105432. [CrossRef]
10. Bonaiti, L.; Concli, F.; Gorla, C.; Rosa, F. Bending fatigue behaviour of 17-4 PH gears produced via selective laser melting. *Procedia Struct. Integr.* **2019**, *24*, 764–774. [CrossRef]
11. Concli, F.; Bonaiti, L.; Gerosa, R.; Cortese, L.; Nalli, F.; Rosa, F.; Gorla, C. Bending Fatigue Behavior of 17-4 PH Gears Produced by Additive Manufacturing. *Appl. Sci.* **2021**, *11*, 3019. [CrossRef]
12. Olsson, E.; Olander, A.; Öberg, M. Fatigue of gears in the finite life regime—Experiments and probabilistic modelling. *Eng. Fail. Anal.* **2016**, *62*, 276–286. [CrossRef]
13. Lambert, R.; Aylott, C.; Shaw, B. Evaluation of bending fatigue strength in automotive gear steel subjected to shot peening techniques. *Procedia Struct. Integr.* **2018**, *13*, 1855–1860. [CrossRef]
14. Zhang, J.; Zhang, Q.; Xu, Z.-Z.; Shin, G.-S.; Lyu, S. A study on the evaluation of bending fatigue strength for 20CrMoH gear. *Int. J. Precis. Eng. Manuf.* **2013**, *14*, 1339–1343. [CrossRef]
15. Lisle, T.J.; Little, C.P.; Aylott, C.J.; Shaw, B.A. Bending fatigue strength of aerospace quality gear steels at ambient and elevated temperatures. *Int. J. Fatigue* **2022**, *164*, 107125. [CrossRef]
16. Dengo, C.; Meneghetti, G.; Dabalà, M. Experimental analysis of bending fatigue strength of plain and notched case-hardened gear steels. *Int. J. Fatigue* **2015**, *80*, 145–161. [CrossRef]

17. Meneghetti, G.; Dengo, C.; Conte, F.L. Bending fatigue design of case-hardened gears based on test specimens. *Proc. Inst. Mech. Eng. Part C J. Mech. Eng. Sci.* **2017**, *232*, 1953–1969. [CrossRef]
18. John, J.; Li, K.; Li, H. Fatigue Performance and Residual Stress of Carburized Gear Steels Part I: Residual Stress. *SAE Int. J. Mater. Manuf.* **2008**, *1*, 718–724. [CrossRef]
19. Dowling, W.E.; Donlon, W.T.; Copple, W.B.; Chernenkoff, R.A.; Darragh, C. Bending Fatigue Behavior of Carburized Gear Steels: Four-Point Bend Test Development and Evaluation. *J. Passeng. Cars* **1996**, *105*, 1330–1339. Available online: https://www.jstor.org/stable/44720852 (accessed on 1 August 2022).
20. Spice, J.J.; Matlock, D.K.; Fett, G. Optimized Carburized Steel Fatigue Performance as Assessed with Gear and Modified Brugger Fatigue Tests. *J. Mater. Manuf.* **2002**, *111*, 589–597. Available online: https://www.jstor.org/stable/44718684 (accessed on 1 August 2022).
21. Kuhn, H.; Medlin, D.; Osman, T.M.; Rigney, J.D.; Weaver, M.; Stevenson, M.; Kalpakjian, S.; House, J.W.; Gillis, P.P.; Robinson, R.M.; et al. Mechanical Testing of Gears. In *ASM Handbook*; ASM International: Materials Park, OH, USA, 2000; Volume 8, pp. 861–872. [CrossRef]
22. Rettig, H. Ermittlung von Zahnfußfestigkeitskennwerten auf Verspannungsprüfständen und Pulsatoren-Vergleich der Prüfverfahren und der gewonnenen Kennwerte. *Antriebstechnik* **1987**, *26*, 51–55.
23. Stahl, K.; Michaelis, K.; Höhn, B.-R. *FVA Research Project 304, Final Report No. 580—Life Time Statistics—Statistical Methods for Estimation of the Machine Element Life Time and Reliability, and Exemplary Application on Gears (German: Lebensdauerstatistik—Statistische Methoden zur Beurteilung von Bauteillebensdauer und Zuverlässigkeit und ihre beispielhafte Anwendung auf Zahnräder)*; FVA-Forschungsheft: Frankfurt am Main, Germany, 1999.
24. Hein, M.; Geitner, M.; Tobie, T.; Stahl, K.; Pinnekamp, B. Reliability of gears—Determination of statistically validated material strength numbers. In Proceedings of the American Gear Manufacturers Association Fall Technical Meeting 2018, Oak Brook, IL, USA, 24–26 September 2018.
25. Tobie, T.; Matt, P. *FVA Guideline 563 I—Recommendations for the Standardization of Load Capacity Tests on Hardened and Tempered Cylindrical Gears*; FVA: Frankfurt am Main, Germany, 2012.
26. Hein, M. Zur Ganzheitlichen Betriebsfesten Auslegung und Prüfung von Getriebezahnrädern. Ph.D. Dissertation, Tecnhical university of Munich, Munic, Germany, 2018.
27. Bonaiti, L.; Gorla, C. Estimation of gear SN curve for tooth root bending fatigue by means of maximum likelihood method and statistic of extremes. *Int. J. Fatigue* **2021**, *153*, 106451. [CrossRef]
28. Bonaiti, L.; Rosa, F.; Rao, P.M.; Concli, F.; Gorla, C. Gear root bending strength: Statistical treatment of Single Tooth Bending Fatigue tests results. *Forsch. Ing.* **2021**, *86*, 251–258. [CrossRef]
29. Seabrook, J.B.; Dudley, D.W. Results of a Fifteen-Year Program of Flexural Fatigue Testing of Gear Teeth. *J. Eng. Ind.* **1964**, *86*, 221–237. [CrossRef]
30. Akata, E.; Altınbalık, M.; Çan, Y. Three point load application in single tooth bending fatigue test for evaluation of gear blank manufacturing methods. *Int. J. Fatigue* **2004**, *26*, 785–789. [CrossRef]
31. Buenneke, R.W.; Slane, M.B.; Dunham, C.R.; Semenek, M.P.; Shea, M.M.; Tripp, J.E. Gear Single Tooth Bending Fatigue Test. *Transactions* **1982**, *91*, 3266–3274. Available online: https://about.jstor.org/terms (accessed on 5 August 2022).
32. J1619_201712; Single Tooth Gear Bending Fatigue Test. SAE International: Warrendale, PA, USA, 2017. Available online: https://www.sae.org/standards/content/j1619_201712/ (accessed on 5 August 2022).
33. Gorla, C.; Rosa, F.; Conrado, E.; Albertini, H. Bending and contact fatigue strength of innovative steels for large gears. *Proc. Inst. Mech. Eng. Part C J. Mech. Eng. Sci.* **2014**, *228*, 2469–2482. [CrossRef]
34. Wildhaber, E. Measuring tooth thickness of involute gears. *Amer. Mach.* **1923**, *59*, 551.
35. Test rigs. Available online: https://www.ifa.ruhr-uni-bochum.de/ifa/dienste/pruefstaende.html.en (accessed on 21 October 2022).
36. Fuchs, D.; Schurer, S.; Tobie, T.; Stahl, K. On the determination of the bending fatigue strength in and above the very high cycle fatigue regime of shot-peened gears. *Forsch. im Ingenieurwesen* **2021**, *86*, 81–92. [CrossRef]
37. Hong, I.; Teaford, Z.; Kahraman, A. A comparison of gear tooth bending fatigue lives from single tooth bending and rotating gear tests. *Forsch. Ing.* **2022**, *86*, 259–271. [CrossRef]
38. Vučković, K.; Čular, I.; Mašović, R.; Galić, I.; Žeželj, D. Numerical model for bending fatigue life estimation of carburized spur gears with consideration of the adjacent tooth effect. *Int. J. Fatigue* **2021**, *153*, 106515. [CrossRef]
39. Rao, S.B.; Schwanger, V.; McPherson, D.R.; Rudd, C. Measurement and Validation of Dynamic Bending Stresses in Spur Gear Teeth. In Proceedings of the ASME International Design Engineering Technical Conferences and Computers and Information in Engineering Conference—DETC2005, Long Beach, CA, USA, 24–28 September 2005; pp. 755–764. [CrossRef]
40. Winkler, K.; Schurer, S.; Tobie, T.; Stahl, K. Investigations on the tooth root bending strength and the fatigue fracture characteristics of case-carburized and shot-peened gears of different sizes. *Proc. Inst. Mech. Eng. Part C J. Mech. Eng. Sci.* **2019**, *233*, 7338–7349. [CrossRef]
41. Fuchs, D.; Schurer, S.; Tobie, T.; Stahl, K. New Consideration of Non-Metallic Inclusions Calculating Local Tooth Root Load Carrying Capacity of High-Strength, High-Quality Steel Gears. Available online: www.geartechnology.com (accessed on 13 July 2022).
42. Fuchs, D.; Schurer, S.; Tobie, T.; Stahl, K. A model approach for considering nonmetallic inclusions in the calculation of the local tooth root load-carrying capacity of high-strength gears made of high-quality steels. *Proc. Inst. Mech. Eng. Part C J. Mech. Eng. Sci.* **2019**, *233*, 7309–7317. [CrossRef]

43. Fuchs, D.; Schurer, S.; Tobie, T.; Stahl, K. Investigations into non-metallic inclusion crack area characteristics relevant for tooth root fracture damages of case carburised and shot-peened high strength gears of different sizes made of high-quality steels. *Forsch. Ing.* **2019**, *83*, 579–587. [CrossRef]

44. Fuchs, D.; Rommel, S.; Tobie, T.; Stahl, K. Fracture analysis of fisheye failures in the tooth root fillet of high-strength gears made out of ultra-clean gear steels. *Forsch. Ing.* **2021**, *85*, 1109–1125. [CrossRef]

45. Mcpherson, D.R.; Rao, S.B. Methodology for translating single-tooth bending fatigue data to be comparable to running gear data. *Gear Technol.* **2008**, *6*, 42–51.

46. Rao, S.B.; Mcpherson, D.H. Experimental characterization of bending fatigue strength in gear teeth. *Gear Technol.* **2003**, *20*, 25–32.

47. Li, M.; Xie, L.-Y.; Li, H.-Y.; Ren, J.-G. Life Distribution Transformation Model of Planetary Gear System. *Chin. J. Mech. Eng.* **2018**, *31*, 24. [CrossRef]

48. Concli, F.; Maccioni, L.; Bonaiti, L. Reliable gear design: Translation of the results of single tooth bending fatigue tests through the combination of numerical simulations and fatigue criteria. *WIT Trans. Eng. Sci.* **2021**, *130*, 111–122. [CrossRef]

49. Concli, F.; Fraccaroli, L.; Maccioni, L. Gear Root Bending Strength: A New Multiaxial Approach to Translate the Results of Single Tooth Bending Fatigue Tests to Meshing Gears. *Metals* **2021**, *11*, 863. [CrossRef]

50. Concli, F.; Maccioni, L.; Fraccaroli, L.; Bonaiti, L. Early Crack Propagation in Single Tooth Bending Fatigue: Combination of Finite Element Analysis and Critical-Planes Fatigue Criteria. *Metals* **2021**, *11*, 1871. [CrossRef]

51. Bonaiti, L.; Bayoumi, A.B.M.; Concli, F.; Rosa, F.; Gorla, C. Gear Root Bending Strength: A Comparison Between Single Tooth Bending Fatigue Tests and Meshing Gears. *J. Mech. Des.* **2021**, *143*, 1–17. [CrossRef]

52. Petersen; Link, R.; Pascual, F.; Meeker, W. Analysis of Fatigue Data with Runouts Based on a Model with Nonconstant Standard Deviation and a Fatigue Limit Parameter. *J. Test. Evaluation* **1997**, *25*, 292. [CrossRef]

53. Mao, T.; Liu, H.; Wei, P.; Chen, D.; Zhang, P.; Liu, G. An improved estimation method of gear fatigue strength based on sample expansion and standard deviation correction. *Int. J. Fatigue* **2022**, *161*. [CrossRef]

54. Dixon, W.J.; Mood, A.M. A Method for Obtaining and Analyzing Sensitivity Data. *J. Am. Stat. Assoc.* **1948**, *43*, 109–126. [CrossRef]

55. Alnahlaui, A.; Tenberge, P. Improved Method for the Determination of Tooth Root Endurance Strength. In Proceedings of the International Conference on Gears 2022, Munich, Germany, 12–14 September 2022.

56. Geitner, M.; Tobie, T.; Stahl, K. *FVA research project 610 IV, final report no. 1432—Materials 4.0—Comprehensive statistical data analysis to evaluate the influence of material and heat treatment properties on the load carrying capacity of gears (German: Werkstoffe 4.0- Erweiterte Datenanalyse zur Bewertung des Einflusses von Werkstoff- und Wärmebehandlungseigenschaften auf die Zahnradtragfähigkeit)*; FVA-Forschungsheft: Frankfurt am Main, Germany, 2021.

57. Hück, M. Ein verbessertes Verfahren für die Auswertung von Treppenstufenversuchen. *Mater. Werkst.* **1983**, *14*, 406–417. [CrossRef]

58. Hösel, T.; Joachim, F. Zahnflankenwälzfestigkeit unter Berücksichtigung der Ausfallwahrscheinlichkeit. *Antriebstechnik* **1978**, *17*.

59. Rossow, E. Eine einfache Rechenschiebernäherung an die den normal scores entsprechenden Prozentpunkte. *Qualitätskontrolle* **1964**, *9*, 146–147.

60. Spindel, J.; Haibach, E. *Some Considerations in the Statistical Determination of the Shape of S-N Curves*; ASTM International: West Conshohocken, PA, USA, 1981; p. 89. [CrossRef]

61. Beretta, S. *Affidabilità Delle Costruzioni Meccaniche*; Springer: Milan, Italy, 2009. [CrossRef]

62. Urbano, M.F.; Cadelli, A.; Sczerzenie, F.; Luccarelli, P.; Beretta, S.; Coda, A. Inclusions Size-based Fatigue Life Prediction Model of NiTi Alloy for Biomedical Applications. *Shape Mem. Superelasticity* **2015**, *1*, 240–251. [CrossRef]

63. Nelson, W. *Applied Life Data Analysis*; Wiley: Hoboken, NJ, USA, 1982. [CrossRef]

64. Nelson, W. *Accelerated Testing: Statistical Models, Test Plans and Data Analyses*; John Wiley & Sons: Hoboken, NJ, USA, 2004; p. 601.

65. Winkler, K.J.; Tobie, T.; Stahl, K. Influence of grinding zones on the tooth root bending strength of case carburized gears. *Forsch. Ing.* **2021**, *86*, 661–671. [CrossRef]

66. Gasparini, G.; Mariani, U.; Gorla, C.; Filippini, M.; Rosa, F. Bending fatigue tests of helicopter case carburized gears: Influence of material, design and manufacturing parameters. *Gear Technol.* **2009**, 68–76. Available online: www.geartechnology.com (accessed on 22 October 2022).

67. Gorla, C.; Rosa, F.; Conrado, E. Bending Fatigue Strength of Case Carburized and Nitrided Gear Steels for Aeronautical Applications. *Int. J. Appl. Eng. Res.* **2017**, *12*, 11306–11322. Available online: http://www.ripublication.com (accessed on 2 August 2022).

68. Castillo, E.; Mínguez, R.; Conejo, A.J.; Pérez, B.; Fontenla, O. Estimating the parameters of a fatigue model using Benders' decomposition. *Ann. Oper. Res.* **2011**, *210*, 309–331. [CrossRef]

69. Pascual, F.G.; Meeker, W.Q. Estimating Fatigue Curves With the Random Fatigue-Limit Model. *Technometrics* **1999**, *41*, 277–289. [CrossRef]

70. Lorén, S. Estimating fatigue limit distributions under inhomogeneous stress conditions. *Int. J. Fatigue* **2004**, *26*, 1197–1205. [CrossRef]

71. Castillo, E.; Fernández-Canteli, A. A compatible regression Weibull model for the description of the three-dimensional fatigue σM–N–R field as a basis for cumulative damage approach. *Int. J. Fatigue* **2021**, *155*, 106596. [CrossRef]

72. Canteli, A.F.; Castillo, E.; Blasón, S.; Correia, J.; de Jesus, A. Generalization of the Weibull probabilistic compatible model to assess fatigue data into three domains: LCF, HCF and VHCF. *Int. J. Fatigue* **2022**, *159*, 106771. [CrossRef]

73. Freudenthal, A.M.; Gumbel, E.J. Physical and Statistical Aspects of Fatigue. *Adv. Appl. Mech.* **1956**, *4*, 117–158. [CrossRef]
74. Beretta, S.; Clerici, P.; Matteazzi, S. The effect of sample size on the confidence of endurance fatigue tests. *Fatigue Fract. Eng. Mater. Struct.* **1995**, *18*, 129–139. [CrossRef]
75. Loren, S. Fatigue limit estimated using finite lives. *Fatigue Fract. Eng. Mater. Struct.* **2003**, *26*, 757–766. [CrossRef]
76. Lorén, S.; Lundström, M. Modelling curved S–N curves. *Fatigue Fract. Eng. Mater. Struct.* **2005**, *28*, 437–443. [CrossRef]
77. Pascual, F.G. Theory for Optimal Test Plans for the Random Fatigue-Limit Model. *Technometrics* **2003**, *45*, 130–141. [CrossRef]
78. Miller, R.; Nelson, W. Optimum Simple Step-Stress Plans for Accelerated Life Testing. *IEEE Trans. Reliab.* **1983**, *R-32*, 59–65. [CrossRef]
79. Horstman, R.; Peters, K.; Gebremedhin, S.; Meltzer, R.; Vieth, M.B.; Nelson, W. Fitting of Fatigue Curves with Nonconstant Standard Deviation to Data with Runouts. *J. Test. Eval.* **1984**, *12*, 69–77. [CrossRef]
80. Mitra, E.D.; Hlavacek, W.S. Parameter estimation and uncertainty quantification for systems biology models. *Curr. Opin. Syst. Biol.* **2019**, *18*, 9–18. [CrossRef]
81. Meeker, W.Q.; Escobar, L.A. Teaching about Approximate Confidence Regions Based on Maximum Likelihood Estimation. *Am. Stat.* **1995**, *49*. [CrossRef]
82. Wiel, S.V.; Meeker, W. Accuracy of approx confidence bounds using censored Weibull regression data from accelerated life tests. *IEEE Trans. Reliab.* **1990**, *39*, 346–351. [CrossRef]

*Article*

# Shape Deviation Network of an Injection-Molded Gear: Visualization of the Effect of Gate Position on Helix Deviation

**Jing Chong Low *, Daisuke Iba, Daisuke Yamazaki and Yuichiro Seo**

Precision Manufacturing Laboratory, Department of Mechanical Engineering, Kyoto Institute of Technology, Goshokaidou-cho, Matsugasaki, Sakyo-ku, Kyoto 606-0951, Japan; iba@kit.ac.jp (D.I.); daisuke.yama.0912@gmail.com (D.Y.); seoyuu1210@gmail.com (Y.S.)

* Correspondence: jingchong1998@gmail.com; Tel.: +81-80-9633-0166

**Abstract:** The purpose of this study is to develop an evaluation method for assessing the relative relationship between gear tooth shape deviations on every gear tooth using network theory. Our previous study introduced a method for representing the phase difference between each helix deviation as a network, demonstrating that it is possible to identify the relative relationship between gear tooth deviations. However, there has been no in-depth analysis of the impact of injection molding on the gear's phase difference network. In this paper, we begin by measuring the gear tooth shape deviation, calculating the correlation coefficient, and expressing it as an adjacency matrix. When the adjacency matrix was visualized and displayed as a pixel plot, a periodic pattern was formed. The relationship between the position of the gate used to inject molten resin material into the injection mold and the pixel plot were then investigated in detail, and it was confirmed that the helix deviation network of a gear manufactured with the injection molding process is useful as a new indicator for the manufacturing error of injection-molded plastic gear.

**Keywords:** gear inspection; injection-molded plastic gear; helix deviation; network theory; correlation coefficient; pixel plot; network image

**Citation:** Low, J.C.; Iba, D.; Yamazaki, D.; Seo, Y. Shape Deviation Network of an Injection-Molded Gear: Visualization of the Effect of Gate Position on Helix Deviation. *Appl. Sci.* **2024**, *14*, 2013. https://doi.org/10.3390/app14052013

Academic Editors: Aleksandar Miltenovic, Franco Concli and Stefan Schumann

Received: 31 January 2024
Revised: 23 February 2024
Accepted: 25 February 2024
Published: 29 February 2024

## 1. Introduction

Gears play an important role in power transmission mechanisms and are widely used across various industries. With the growing demand for high-volume production, there is a need to manufacture transmission gears that are both highly efficient and low in noise. The properties of gear tooth flanks generally affect not only the strength but also the efficiency and noise level of the gears in power transmissions [1–5]. While high precision grade gear pairs are believed to produce low-noise and low-vibration transmission systems, there are instances where even with the use of such gears, quiet gear power transmissions cannot be achieved [6–8]. This suggests that the current gear accuracy evaluation system may still be insufficient. The existing gear inspection parameters only account for tooth profile deviation and tooth helix deviation of individual teeth, without considering the relative evaluation among the deviations that may cause unnecessary vibration during power transmission [9,10].

Network theory is an analytical approach that is commonly employed to analyze complex system structures, such as railway networks, airline networks, and the network of spreading of COVID-19, as well as social network analysis of attitudes towards immigrants [11–13]. In addition, the method can be utilized to understand and elucidate the deviation of manufactured gears, for which Iba et al. proposed a novel approach based on network theory for expressing the relationship between tooth helix deviation curves in helical gears [14,15]. The study is accomplished by calculating the inner product between two helix deviations, which generates correlation coefficients which is an indicator capable of acquiring the phase difference between the deviations in the direction of the helix. Based on this approach, a network image of the gear is constructed, with each tooth of the gear representing a node and the correlation coefficient representing a link, respectively. Consequently, a comprehensive overview of the entire helix deviation curve of a gear is obtained, a feat that was previously challenging to achieve using conventional chart diagrams of helix deviation. However, to date, no detailed studies have created a tooth helix deviation network using deviation data of plastic gear manufactured through injection molding, nor have any studies investigated the impact of the manufacturing method on the phase difference network of the gear.

In this paper, this study aims to develop an evaluation method to assess the relative relationship between shape deviations on each tooth of an injection-molded gear by utilizing network theory. Firstly, the gear tooth shape deviation on each tooth of the injection-molded plastic gear was measured, and the data acquired were used to construct a helix deviation network. Subsequently, the adjacency matrix of the network was output as an image in the form of a pixel plot. The periodic pattern that formed in the pixel plot of the helix deviation network was then meticulously investigated by generating network images consisting of various fixed correlation coefficient ranges.

## 2. Materials and Methods

The present study presents a specialized approach to extract and visualize the periodicity imprinted on the tooth flank during the gear manufacturing process. The approach is based on visualizing the network between helix deviations, as suggested and devised by Iba et al. [14,15]. The first section of the paper provides a detailed description of the gear specification, followed by an overview of the adopted analysis method. The approach includes preprocessing the helix deviation data obtained from the gear inspection machine, computation of the correlation coefficient between two helix deviations, generation of the adjacency matrix, and visualization of the helix deviation network of injection-molded plastic gear. The methodology provides an effective way to extract and visualize the periodicity information left on the gear tooth flank during the gear manufacturing process.

### 2.1. Gear Specification

Figure 1a shows the gear used in this study, an injection-molded plastic gear with a module of 1.0. Made from polyoxymethylene (POM), it is also referred to as acetal or polyacetal, and the gear has a total of 48 teeth. The gear's dimensions are displayed in Figure 1b, with its specifications outlined in Table 1 below. To accurately assess the deviation network of the helix in gears manufactured via injection molding, it is worth noting that the analysis in this study was performed on gears that were not subjected to any driving tests prior to the analysis.

**Table 1.** Gear specification.

| Parameter | Value (Target Gear) |
| --- | --- |
| Module (mm) | 1.0 |
| Number of Teeth (−) | 48 |
| Helix Angle (deg.) | 0 |
| Pressure Angle (deg.) | 20 |
| Profile Shift Coefficient (−) | 0 |
| Face width (mm) | 8.0 |
| Tip Diameter (mm) | 50.0 |
| Root Diameter (mm) | 45.5 |
| Material | Polyoxymethylene (POM) |

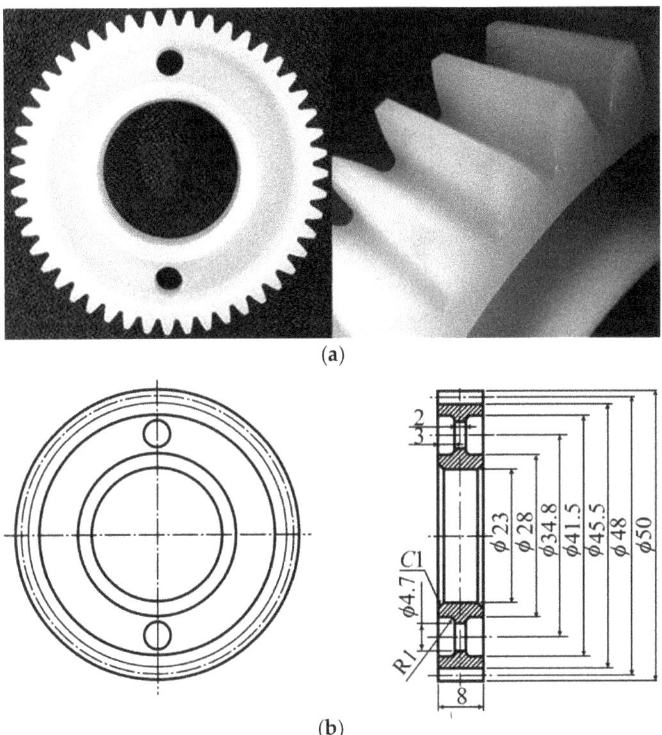

Figure 1. (a). Injection-molded plastic gear used in this study. (b). Measurement of gear used in this study.

## 2.2. Gear Inspection

To analyze the helix deviation curves of the injection-molded plastic gear, we adopted a method proposed and devised by Iba et al. elaborated in reference article number 14th and 15th. Firstly, the helix deviation data of the gear were obtained by using a gear measuring machine. In our study, we employed a Computer Numerical Control (CNC) automatic gear inspection machine (model name: CLP-35) by TPR OSAKA SEIMITSU KIKAI Co., Ltd., Higashi-Osaka, Osaka, Japan to carry out the gear inspection process to obtain the data on helix deviation. Figure 2 shows an image of an injection-molded plastic gear undergoing the gear inspection process with the utilization of the gear inspection machine CLP-35. The specification of the gear inspection machine is shown in Table 2. When the gear inspection process is completed, the obtained data are converted and stored in the form of a Microsoft Excel 2019 file. The file is then imported to "MATLAB R2023a" programming software for further detailed analysis. For instance, the analyses executed in this study consisted of preprocessing of helix deviation data obtained from a gear inspection machine, computation of the correlation coefficient between two helix deviations, generation of the adjacency matrix, and visualization of the helix deviation network of injection-molded plastic gear.

**Figure 2.** Injection-molded plastic gear undergoing inspection process.

**Table 2.** Specification of Gear Measuring Machine.

| Parameter | Value (Gear Measuring Machine) |
| --- | --- |
| Normal Module (mm) | 0.5~12 |
| Number of Teeth (−) | 10~500 |
| Gear Outer Diameter (mm) | Max $\varphi$350 |
| Basic Circle Diameter (mm) | 0~$\varphi$300 |
| Tooth Width (mm) | Max 400 |
| Tangent Length for Profile Measurement (mm) | ±120 |
| Helix Angle (deg.) | 0~±60 |
| Gear Shaft Length (mm) | 50~400 |
| Gear Weight (kg) | Max 50 |
| Resolution (V) | 0.0001 |
| Power Supply (V) | AC 100 |
| Capacitance (KVA) | 2 |
| Machine Weight (kg) | 1500 |
| Dimension (W × D × H) (mm) | 1133 × 1071 × 1995 |

### 2.3. Data Preprocessing

#### 2.3.1. Elimination of Direct Current (DC) Component

When addressing tooth helix deviation, a crucial factor to consider is the direct current (DC) component or average value of the waveform. Removing the DC component is prioritized as it has a significant impact on pitch deviation but does not affect the phase shift of the helix deviation. Following this, the helix deviation curve after removing the DC component for $i^{th}$ tooth number, $f_{\beta i}(x)$, is calculated with the following Equation (1) in this paper.

$$f_{\beta i}(x) = \widehat{f_{\beta i}}(x) - f_{\beta\,DC\,i}(x) \tag{1}$$

The helix deviation curve obtained from the data of gear inspection for $i^{th}$ tooth number is regarded as $\widehat{f_{\beta i}}(x)$, and the DC component or average value of the curve for $i^{th}$ tooth number is defined as $f_{\beta\,DC\,i}(x)$. Additionally, $x$ denotes the position of the evaluation length of helix deviation.

#### 2.3.2. Elimination of Slope Component

The inclusion of a slope component in the measurement signal such as the signal of helix deviation of this study can pose a challenge when attempting to extract its periodicity through correlation function analysis. Therefore, it is crucial to eliminate the gradient component to derive the exact periodicity of the signal. In this paper, the helix deviation curve after the removal of DC and slope component for the $i^{th}$ tooth number, $f_{H\beta i}(x)$, is calculated with the following Equation (2).

$$f_{H\beta i}(x) = f_{\beta i}(x) - y_{H\beta i}(x) \tag{2}$$

$f_{\beta i}(x)$ is regarded as the helix deviation curve after the DC component for $i^{th}$ tooth number is eliminated, and $y_{H\beta i}(x)$ is regarded as the slope component of $f_{\beta i}(x)$ calculated by utilizing the least square method.

### 2.3.3. Elimination of Low-Frequency Component

It is crucial to acknowledge that the signal of a helix deviation often comprises a waviness component, which pertains to a low-frequency element. As significant low-frequency components of the signal can substantially impact the outcome of the correlation function analysis, it is crucial to consider the elimination of the low-frequency component of the helix deviation signal. Equation (3) illustrates the Fourier transform employed on the helix deviation curve, $f_{H\beta i}(x)$, following the removal of the slope component.

$$F_{H\beta i}(\omega/2\pi) = \int_{-\infty}^{\infty} f_{H\beta i}(x) \cdot e^{-i \cdot \omega x} dx \tag{3}$$

Subsequently, the largest waviness component, $F_{H\beta i \, low}$, of the tooth helix deviation after being Fourier transformed is then eliminated by utilizing the following Equation (4). The Fourier transformed helix deviation signal following the elimination of the largest waviness component is regarded as $F_{\beta i \, eli}$ as illustrated in Equation (4).

$$F_{\beta i \, eli} = F_{H\beta i}(\omega/2\pi) - F_{\beta i \, low} \tag{4}$$

In consequence, the helix deviation after the elimination of the largest waviness component, $F_{\beta i \, high}(x)$, is then derived by applying inverse Fourier transform to the Fourier transformed helix deviation signal following the elimination of the largest waviness component $F_{\beta i \, eli}$ as illustrated in the following Equation (5).

$$F_{\beta i \, high}(x) = \int_{-\infty}^{\infty} F_{\beta i \, eli} \cdot e^{i \cdot \omega x} d\omega \tag{5}$$

### 2.3.4. Gaussian Filter

In order to effectively extract the desired frequency component of the study, a designated filter is utilized in this procedure. In particular, the Gaussian filter is employed to remove the roughness component from the frequency component of the signal. This filter can be employed alongside the Fourier transform and inverse Fourier transform operation to effectively remove low-frequency components. The filter's weight function, as defined in Equation (6), has an amplitude transfer of 50% at the cutoff value of $\lambda_c$.

$$s(x) = \frac{1}{\alpha \lambda_c} e^{-\pi \left(\frac{x}{\alpha \lambda_c}\right)^2} \tag{6}$$

The coefficient $\alpha$ is defined in the following Equation (7).

$$\alpha = \left(\frac{ln2}{\pi}\right)^{0.5} \tag{7}$$

The filtering of the signal is executed by utilizing convolution integration of the measured helix deviation curves and weight functions as shown in the following Equation (8).

$$\hat{f}_{\beta i}(x) = \int \overline{f}_{\beta i}(\tau) s(x - \tau) d\tau \tag{8}$$

### 2.4. Correlation Function

In this paper, a helix deviation network of the gear is generated from the measured helix deviation data according to the method proposed and devised in the previous studies [14,15]. This section describes the derivation method of the shape deviation network. Firstly, nodes and links are required for constructing networks. To obtain a precise understanding of the relationships that exist among nodes within a network, it is important to examine not only the existence of the links within the nodes but also the strength or intensity of their connection. This study has achieved this by computing correlation coefficients between helix deviations. In the methodology adopted based on previous studies, the teeth of gears are regarded as nodes, and the computed correlation coefficients serve as links within the network. The correlation coefficient is determined by deriving the inner product between two helix deviations as illustrated in the following Equation (9).

$$\langle f_{\beta j}(x), f_{\beta k}(x) \rangle = \frac{1}{L} \int_0^L f_{\beta j}(x) f_{\beta k}(x) dx \tag{9}$$

where $f_{\beta j}(x)$ represents the helix deviation of $j^{th}$ tooth. $j$ and $k$ are positive integers from tooth number 1 to the number of teeth $z$; however, $k \neq j$. $x$ is defined as the position of the evaluation length and $L$ is defined as the integral range. The orthogonality of signals can be expressed by defining the inner product as illustrated in Equation (9). When $\langle f_{\beta j}(x), f_{\beta k}(x) \rangle = 0$, $f_{\beta j}(x)$ and $f_{\beta k}(x)$ are orthogonal. By calculating the inner product as illustrated in Equation (9) above, the phase difference between the two tooth helix deviations can be obtained. The quotient of the inner product and the norm of each deviation is calculated as the correlation coefficient. Additionally, the norm of the helix deviation is defined in the following Equation (10).

$$\|f_{\beta j}(x)\| = \sqrt{\frac{1}{L} \int_0^L \{f_{\beta j}(x)\}^2 dx} \tag{10}$$

By applying both Equations (9) and (10), the correlation coefficient of a network is defined as the following Equation (11).

$$r_{j,k} = \frac{\langle f_{\beta j}(x), f_{\beta k}(x) \rangle}{\|f_{\beta j}(x)\| \|f_{\beta k}(x)\|} \tag{11}$$

It is important to note that the correlation coefficient measures the degree of linear relationship between two variables, and it ranges from the value $-1.0$ to $1.0$, where the value $-1.0$ denotes a perfect negative correlation, the value 0 denotes no correlation, and the value 1.0 indicates a perfect positive correlation. When the two tooth helix deviations exhibit a small phase difference, the correlation coefficient tends to approach the value of 1.0. Conversely, when the two deviations display a large phase difference, the correlation coefficient tends to approach the value of $-1.0$. Therefore, it is crucial to consider the phase difference between the two tooth helix deviations when interpreting the value of the correlation coefficient.

*2.5. Adjacency Matrix*

To represent complete information about a network, links in the network must be traced and represented. Therefore, network theory [11] mathematically represents the network by listing links in a matrix called an adjacency matrix, where the adjacency matrix $A$ of a network diagram with $N$ edges have $N$ rows and $N$ columns, and the element $A_{m,n}$ for which edge $(m, n)$ has weight of $w_{m,n}$ is usually defined as the following Equation (12).

$$A_{m,n} = \begin{cases} w_{m,n} \ (If \ v_m \ and \ v_n \ are \ connected) \\ 0 \ (If \ v_m \ and \ v_n \ are \ not \ connected \ to \ each \ other) \end{cases} \tag{12}$$

In this study, the 1st row and the 1st column of the matrix express the number of gear teeth in the matrix. The correlation coefficient among the helix deviations of the gear can be found starting in the 2nd row and 2nd column of the adjacency matrix. Each component from the 2nd row and 2nd column represents the correlation coefficient among the helix deviations of the gear. The diagonal elements of the matrix are defined as 0. By employing the adjacency matrix, it is possible to quantitatively evaluate the helix deviation network of a gear.

*2.6. Visualization*

2.6.1. Pixel Plot

In this section, a diagram consisting of pixels is generated from the adjacency matrix. Similar to the tooth number ordering in the adjacency matrix, the horizontal and vertical axes both represent the number of teeth, with the number of tooth number on the horizontal axis starting from left to right in ascending order, and the number of tooth number on the vertical axis starting from the top to bottom in ascending order. Each tooth number, $z$, possesses a total number of $z - 1$ correlation coefficients with other teeth, and correlation coefficients of all teeth are listed together in an adjacency matrix with a dimension of $z \times z$ excluding the 1st row and the 1st column of the matrix that expresses the number of gear teeth. The degree of the correlation coefficient, $r_{j,k}$ is represented by the color bar positioned on the right side next to the plot, the yellowish color represents a correlation coefficient with a value closer to 1.0. On the contrary, the bluish color represents a correlation coefficient with a value closer to $-1.0$. By applying this method, the relative relationship between each gear teeth can be studied and understood.

2.6.2. Network Image

A diagram consisting of several points and lines connecting the points is called a graph. When a graph consists of weighted edges that incorporate various functions, such as time or distance, it is commonly referred to as a "network". In network theory, the vertices and edges are denoted as nodes and links, respectively. The graph $G$ can be determined by its vertex set $V$ and edge set $E$. Therefore, the expression for the graph $G$ can be represented mathematically by using the following Equation (13).

$$G = (V, E) \tag{13}$$

In this study, the vertex set $V$ consists of the teeth of gears and is expressed by the following equation, where $z$ represents the tooth number.

$$V = \{v_1, v_2, \cdots, v_z\} \tag{14}$$

Next, the edge connected to the $m^{th}$ and $n^{th}$ gear tooth number is expressed as $\{v_m, v_n\}$. Thus, the edge set $E$ is defined as the following equation.

$$E = \{(v_1, v_2), (v_2, v_3), \cdots, (v_{z-1}, v_z)\} \tag{15}$$

In this study, correlation coefficients are used to assign weights to the edges, allowing one to generate a network image that displays the relative relationship between gear teeth within a specific range of correlation coefficients.

## 3. Results

In this section, the analysis method is applied to the injection-molded cylindrical gear shown in Section 2.1, and the result obtained will be presented and discussed. To ensure a thorough evaluation of the results, the section has been divided into several distinct parts comprising the helix deviation curve, adjacency matrix, pixel plot, and network image.

*3.1. Helix Deviation Curve*

3.1.1. Original Data

Figure 3 shows the left helix deviation curve of tooth number 1st plotted with data obtained from the gear inspection process. For ease of identification, the helix deviation plotted with all data points is indicated by the curve in black color, and the helix deviation after the elimination of 20% of data points at both ends is indicated by the curve in red color. Due to the presence of chamfer and rounding of the tooth on the tooth flank, the data at both ends of helix deviation possess inappropriate weighted information that will have an unnecessary influence on the analysis result. Therefore, 20% of data points at both ends of the helix deviation are eliminated to enhance the accuracy of the analysis result.

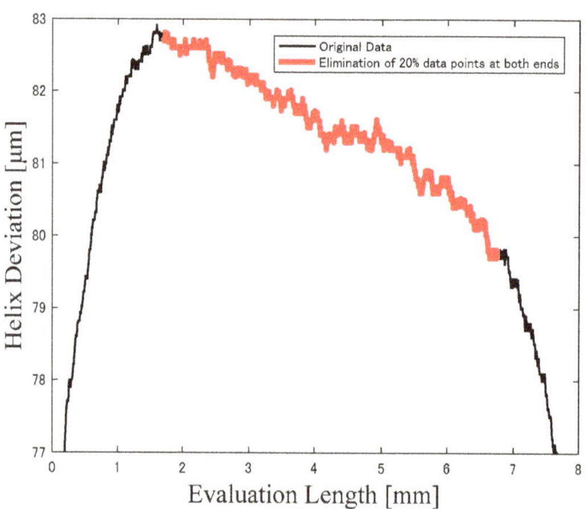

**Figure 3.** Helix deviation of tooth number 1st: helix deviation plotted with all data points is indicated by the color black, and the helix deviation after 20% of data points at both ends is eliminated is indicated by the color red.

The helix deviation data obtained from the gear inspection process are plotted in the same graph, and Figure 4a and Figure 4b show the left and right helix deviation curves of all gear teeth, respectively. The helix deviation curves of tooth number 1st~48th are shown accordingly in the graph from top to bottom. From the results, we observed that left and right helix deviations of the analyzed POM gear exhibit a negative gradient, while the gradient of each tooth varies among each tooth.

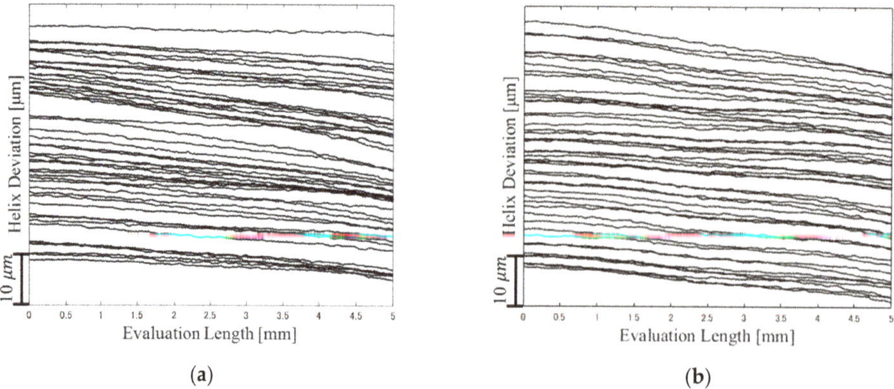

**Figure 4.** Helix deviation curves derived from helix deviation data obtained from the gear inspection process: (**a**) Left helix deviation where the helix deviation curves of tooth number 1st~48th are shown accordingly in the graph from top to bottom; (**b**) Right helix deviation where the helix deviation curves of tooth number 1st~48th are shown accordingly in the graph from top to bottom.

### 3.1.2. Elimination of Direct Current (DC) and Slope Component

The helix deviation curves of left and right gear teeth after the elimination of DC and slope components are shown in Figure 5a and Figure 5b, respectively. In contrast to the helix deviation curves of the original data, both left and right helix deviations display a relatively uniform gradient. Nonetheless, certain curves exhibit a slightly convex shape, with a minor dip in the central region of the curves.

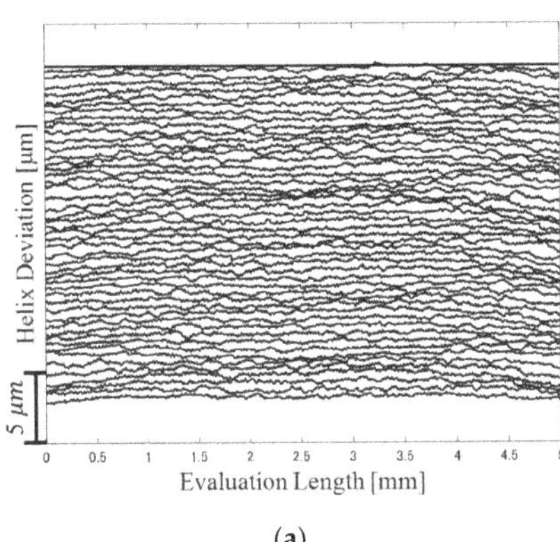

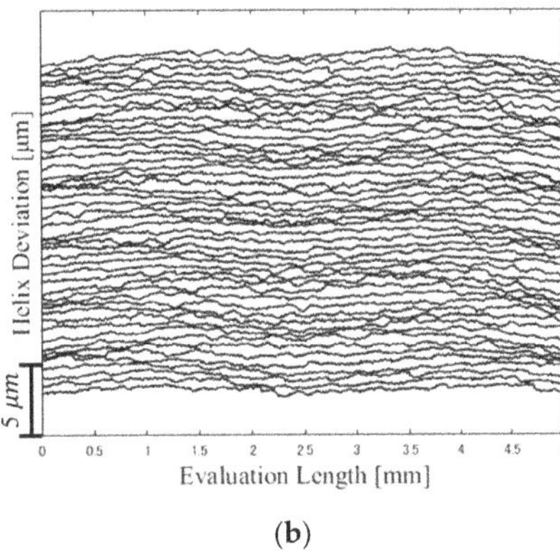

(a)                                               (b)

**Figure 5.** Helix deviation curves after elimination of DC and slope component: (**a**) Left helix deviation where the helix deviation curves of tooth number 1st~48th are shown accordingly in the graph from top to bottom; (**b**) Right helix deviation where the helix deviation curves of tooth number 1st~48th are shown accordingly in the graph from top to bottom.

### 3.2. Adjacency Matrix

### 3.2.1. Adjacency Matrix of Left and Right Helix Deviation

In this section, the correlation coefficient of each tooth is then calculated and listed as an adjacency matrix. In this study, the injection-molded plastic gear used is a gear consisting of 48 teeth. Therefore, the adjacency matrix formed is a 48 × 48 dimensions matrix. Figure 6a shows the adjacency matrix enclosed with the correlation coefficient of left helix deviation. Due to the enormous size of the adjacency matrix, only the information of the tooth numbers 1st to 8th are shown here. To study the network of the helix deviation, we then visualize the adjacency matrix by generating a pixel plot, and a color bar is added to express the strength of the correlation coefficient. Figure 6b shows the pixel plot corresponding to the adjacency matrix shown in Figure 5a. From the pixel plot, the pixels of tooth number pairs 3rd and 4th, 3rd and 5th, and 4th and 5th are filled with the color yellow as the correlation coefficients of the 3 pairs are 0.797, 0.773, and 0.773, respectively. On the contrary, the pixels of tooth number pairs 2nd and 5th and 5th and 7th are filled with the color blue as the correlation coefficients of the pairs are −0.002 and 0.032, respectively. With the visualization of pixel plots, the helix deviation network of gear with an enormous number of teeth can be understood easily.

The adjacency matrix and its visualization allow the expression of the entire tooth helix deviation curve from a bird's-eye view, which is a significant improvement over the conventional chart diagram that organizes the deviation curves tooth by tooth. The conventional chart diagram poses difficulty in expressing an overview of the tooth helix deviation curve, which is addressed by the utilization of this method. By employing this approach, the entire curve can be visualized with ease, providing a comprehensive representation of the network of helix deviation.

|   | 1 | 2 | 3 | 4 | 5 | 6 | 7 | 8 |
|---|---|---|---|---|---|---|---|---|
| 1 | 0.000 | 0.214 | 0.743 | 0.643 | 0.494 | 0.591 | 0.370 | 0.237 |
| 2 | 0.214 | 0.000 | 0.225 | 0.090 | -0.002 | 0.187 | 0.542 | 0.486 |
| 3 | 0.743 | 0.225 | 0.000 | 0.797 | 0.773 | 0.665 | 0.259 | 0.255 |
| 4 | 0.643 | 0.090 | 0.797 | 0.000 | 0.773 | 0.531 | 0.100 | 0.114 |
| 5 | 0.494 | -0.002 | 0.773 | 0.773 | 0.000 | 0.541 | 0.032 | 0.118 |
| 6 | 0.591 | 0.187 | 0.665 | 0.531 | 0.541 | 0.000 | 0.343 | 0.241 |
| 7 | 0.370 | 0.542 | 0.259 | 0.100 | 0.032 | 0.343 | 0.000 | 0.420 |
| 8 | 0.237 | 0.486 | 0.255 | 0.114 | 0.118 | 0.241 | 0.420 | 0.000 |

(**a**)

(**b**)

**Figure 6.** (**a**) Adjacency matrix of tooth number 1st~8th of left helix deviation; (**b**) Pixel plot corresponding to adjacency matrix of tooth number 1st~8th of left helix deviation.

### 3.2.2. Strong Correlation and Weak Correlation

This section shows the curves with the greatest and weakest correlation coefficients between the helix deviation data of the left and right helix deviations of the injection-molded gear. Figure 7a shows the helix deviation curves of tooth number 20th and 45th that exhibit the greatest correlation coefficient in the left helix deviation. The value of the correlation coefficient is 0.892. The helix deviation curve of tooth number 20th and 45th are identical to each other as portrayed in Figure 7a. Despite having slightly different fluctuations in the curves, the phase difference between the curves is almost zero. On the contrary, the minimum value of the correlation coefficient is −0.448, which is exhibited by the pair of tooth number 9th and 27th shown in Figure 7b. As opposed to Figure 7a, the helix deviation curves in Figure 7b possess a large phase difference. By calculating correlation coefficients of helix deviations, we can uncover new insights and obtain phase information between helix deviations of 2 gear teeth.

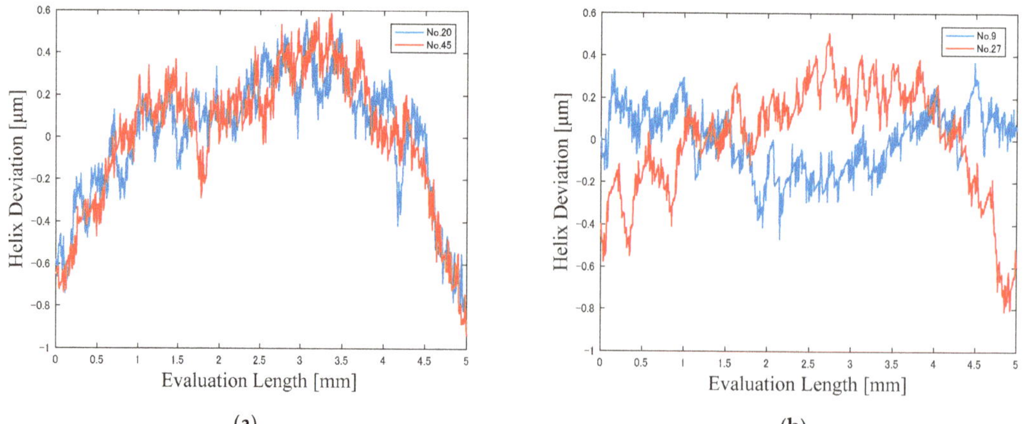

(**a**)

(**b**)

**Figure 7.** Left helix deviation pairs with the greatest and lowest correlation coefficient: (**a**) Left helix deviation pair with highest correlation coefficient, tooth number 20th and 45th; (**b**) Left helix deviation pair with lowest correlation coefficient, tooth number 9th and 27th.

### 3.3. Pixel Plot

Figure 8a,b show the pixel plot of left and right helix deviation generated from the adjacency matrix that encloses information on the correlation coefficient after the DC and slope components are eliminated. Since the greatest correlation coefficients of left and right helix deviation are 0.892 and 0.916 and the weakest correlation coefficients of left and right helix deviation are −0.448 and −0.419, respectively, the range of color bar which indicates the degree of the correlation coefficient in pixel plot is set to −0.5 to 1.0. For both left and right helix deviations, a uniform and periodic pattern is formed. In both cases, for almost every tooth number, there are six regions with relatively high correlation coefficients, which are indicated by the yellowish and yellow color of the plot, and six regions with low negative correlation coefficients which are indicated by the greenish and blue color. In the case of the left tooth helix deviation shown in Figure 8a, a grid-like pattern can be observed. Moreover, the region of tooth number 2nd, 7th~10th, 16th~18th, 41st~42nd possesses a lower correlation coefficient in left helix deviation. In the case of the right helix deviation shown in Figure 8b, a houndstooth pattern was formed. Every tooth possesses six high and low correlation coefficient regions, and the size of the regions is fairly uniform. For the grid-like pattern that formed in the pixel plot of left helix deviation and the houndstooth pattern that formed in right helix deviation, the six high and six low correlation coefficient regions occurred at almost the same distance.

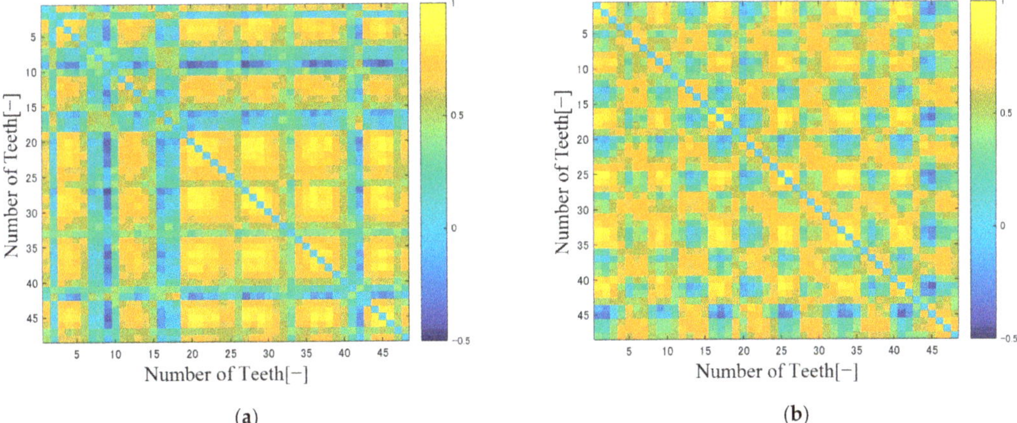

**(a)**  **(b)**

**Figure 8.** Pixel plot: (**a**) Pixel plot of left helix deviation exhibits a grid-like pattern; (**b**) Pixel plot of right helix deviation exhibits a houndstooth pattern.

*3.4. Network Image*

Figures 9 and 10 show the network images of left and right helix deviation generated from the upper triangular of the adjacency matrix that enclosed information on the correlation coefficient after DC and slope components are eliminated. The network images are generated within the correlation coefficient range of −0.5 to 1.0, and a 0.25 interval between each image is set to understand the properties of helix deviations of the gear and avoid overlapping of the links. A total of six network images are generated in descending order with the degree of strength of the correlation coefficients, and the links in red represent the positive correlation coefficient value between two helix deviations, while the links in blue represent the negative correlation coefficient value between two helix deviations. In the network images of both left and right helix deviation, most of the links occurred within the positive ranges, while most of the positive links occurred in the correlation coefficient range of $0.50 \leq r_{j,k} \leq 0.75$.

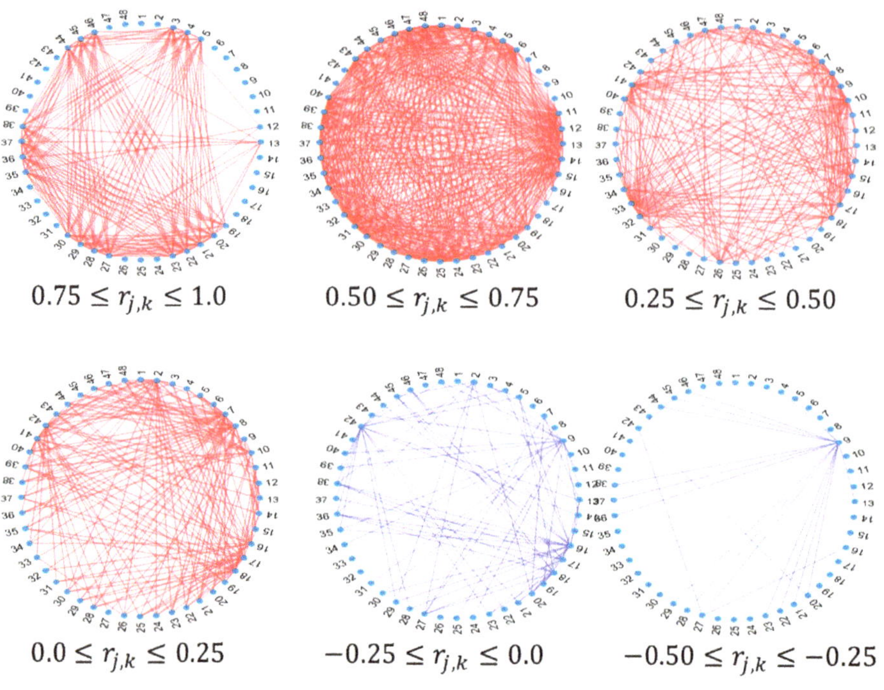

**Figure 9.** Network images of left helix deviation.

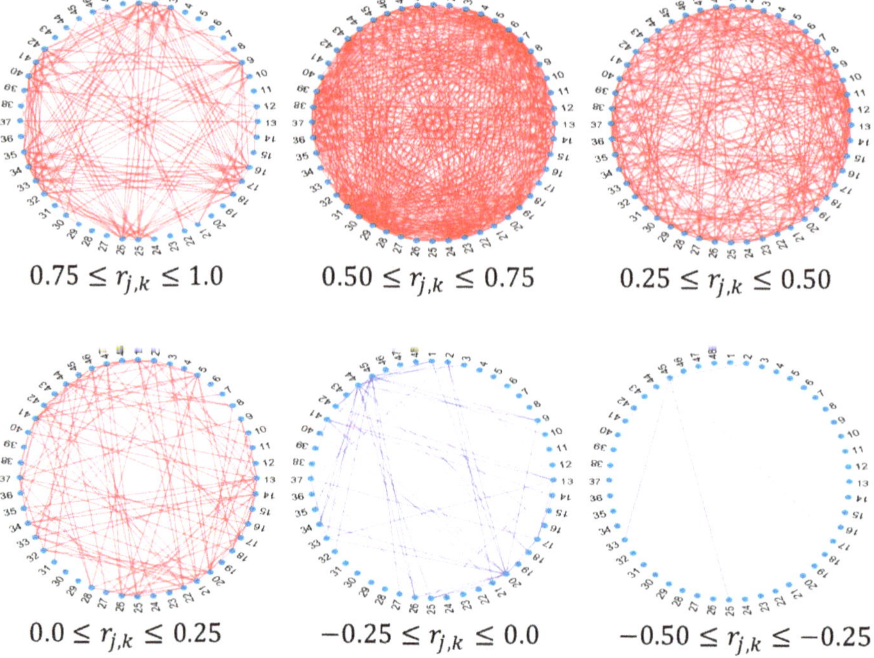

**Figure 10.** Network images of right helix deviation.

Additionally, in both cases of left and right helix deviation, a hexagon-shaped network is formed within the correlation coefficient range of $0.75 \leq r_{j,k} \leq 1.0$. A total of six spots concentrated with links were formed in both the network images. For the left helix deviation shown in Figure 8, the six concentrated link spots are tooth number 3rd~5th, 12th~13th, 19th~23rd, 27th~30th, 35th~38th, 44th~46th, whereas, in right helix deviation shown in Figure 10, the six concentrated link spots are 1st~3rd, 9th~10th, 16th~18th, 24th~26th, 32nd~35th, 39th~42nd.

## 4. Discussion

The present study demonstrates that visualizing the network between helix deviations of an injection-molded plastic gear enables the extraction and visualization of the periodicity left on the gear tooth flank during the gear manufacturing process. To ensure the accurate analysis and extraction of the periodicity left on the gear tooth flank during the injection molding process, the DC and slope components were eliminated to eliminate unnecessary weight that would affect the accuracy of the analysis result. Gaussian filter was not adopted in this study as the extraction of the waviness component would eliminate the periodicity left on the tooth flank of an injection-molded plastic gear.

Furthermore, with the visualization of the pixel plot, the relative relationship between each gear tooth can be recognized easily. When the pixel plots of left and right helix deviation of the injection-molded plastic gear were generated, a grid pattern was formed in the network of the left helix deviation shown in Figure 8a, whereas in the case of the network of right helix deviation shown in Figure 8b, a houndstooth pattern was formed. By comparing the pixel plot of a specific tooth number across the pixel plot vertically or horizontally, the information on the correlation coefficient of the specific tooth with all other 47th teeth could be obtained.

For instance, in the case of the pixel plot of left helix deviation, tooth numbers 7th to 10th were filled with bluish pixels when observed across the plot. This implies that the helix deviations of tooth numbers 7th to 10th exhibit a lower correlation coefficient, meaning that the helix deviations are dissimilar when compared to other teeth. Figure 11a and Figure 11b show the left helix deviation curves of tooth number 3rd~6th and 7th~10th, respectively. When the helix deviation curves of tooth number 7th~10th are compared to the helix deviation curves of tooth number 3rd~6th, which largely possess pixels filled with yellowish color when observed across the plot, it is apparent that the curves have a different distinctive shape. In the case of tooth number 3rd~6th, the helix deviation curves possess convex shape curves, with a slight dent at evaluation length 1.8 mm, whereas in the case of tooth number 7th~10th, the helix deviation curves possess M-shaped curves, where the dent of the curves occurred at evaluation length 2.5 mm.

In the case of pixel plot of right helix deviation, when the pixels of tooth number 1st~3rd are observed horizontally across the plot, pixels filled with yellowish color and bluish color alternated and occurred next to each other, where the yellowish pixels formed first from the left. On the contrary, for tooth number 4th~7th, pixels filled with yellowish and bluish color can be observed across the plot, where the bluish pixels formed first from the left. Figure 12a and Figure 12b show the right helix deviation curves of tooth numbers 1st~3rd and 4th~7th, respectively. When the helix deviation curves of tooth number 1st~3rd are compared to tooth number 4th~7th, helix deviation curves in both cases possess m-shaped curves, where the valley of the curves occurred at evaluation length 2.5 mm. However, the minima of the valley of the curves in tooth number 4th~7th has a smaller value of helix deviation compared to the valley in the case of tooth number 1st~3rd.

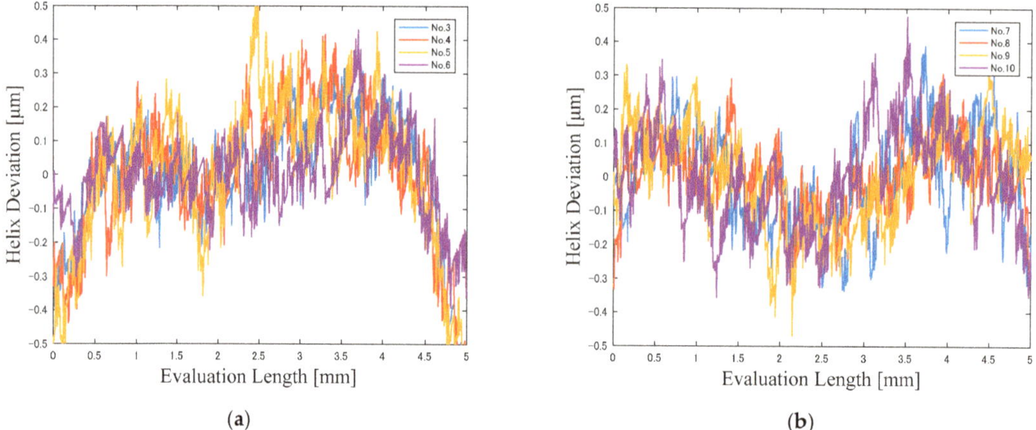

**Figure 11.** Comparison of left helix deviation curves of high and low correlation coefficient regions in the pixel plot: (**a**) Left helix deviation curves of tooth number 3rd~6th; (**b**) Left helix deviation curves of tooth number 7th~10th.

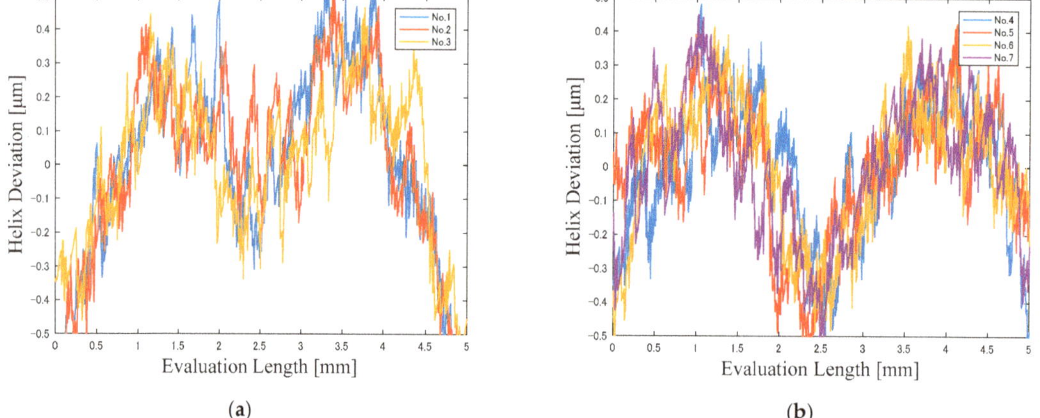

**Figure 12.** Comparison of right helix deviation curves of high and low correlation coefficient regions in the pixel plot: (**a**) Right helix deviation curves of tooth number 1st~3rd; (**b**) Right helix deviation curves of tooth number 4th~7th.

The network of the helix deviations was then output in the form of a network image to investigate the relative relationship between each gear tooth when the correlation coefficient is at a fixed range. In this study, the range was set to $-0.5$~$1.0$ with an interval of $0.25$. When the network of left and right helix deviation was output in the form of network images as shown in Figures 9 and 10, a network pattern with a hexagon shape and six spots of concentrated link spots were formed within the correlation coefficient range of $0.75 \leq r_{j,k} \leq 1.0$. Moreover, these concentrated link spots occurred at the same span. The center of each concentrated link region is approximately eight teeth number next to each other, which correlates with the result in the pixel plot where the center of each high correlation coefficient region is approximately eight teeth number next to the center of the high correlation coefficient region next to it.

When the network images with a coefficient range of $0.75 \leq r_{j,k} \leq 1.0$ were plotted on the illustration of gear used in this study, it became apparent that the injection molding gate has a significant impact on the result of our studies. In plastic injection molding, the gate is the point where molten plastic is injected into the mold during the process. The gate's position plays a pivotal role in the mold design and directly impacts the quality of the molded product. The position of the gate determines the flow of plastic into the mold cavity, and a poorly positioned gate can cause common defects such as shrinkage and weld lines. However, by adjusting the gate position, it is possible to control these defects and enhance the final product's quality significantly [16,17]. Therefore, identifying the optimal gate position is crucial for attaining optimal injection molding outcomes.

In the case of injection-molded gear used in this study, there are six gate marks on the surface of the gear. The injection molding gate mark of the gear used in this study is shown in Figure 13a, and Figure 13b shows the gate mark observed with an electronic microscope. These gate marks are located at the position of tooth numbers 7th, 15th, 23rd, 31st, 39th, and 47th, where the location of each gate mark is eight teeth apart from each other. The concentrated link spots are formed at the region between the gate mark and the weld line. The gate marks are marked with a grey circular ring on the gear illustration and the weld lines are marked with dotted lines in Figure 14. In the case of left helix deviation shown in Figure 14a, the concentrated link spots, highlighted by the large blue circles, were formed on the left side of the gate mark, whereas in the case of right helix deviation shown in Figure 14b, the concentrated link spots were formed on the right side of the gate mark.

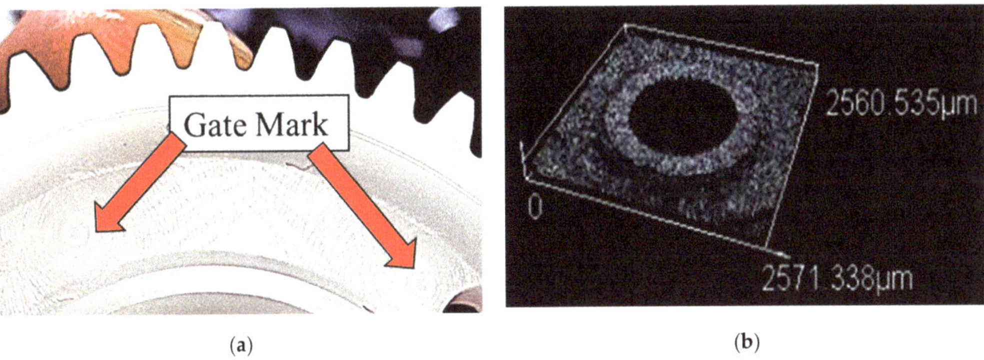


(a)                                                                 (b)

**Figure 13.** Gate mark of injection-molded plastic gear used in this study: (**a**) Gate mark of gear; (**b**) Gate mark observed with electronic microscope.

In the plastic injection molding process, fountain flow occurs when the material at the flow front is pushed forward and channeled out of the injection molding gate. During the mold filling process, the material is injected into an empty melt flow channel fountains to the channel walls. The fountain flow creates similar left helix deviations at the left side of the gate mark to the weld line to its left, and similar right helix deviations at the right side of the gate mark to the weld line to its right. Since the temperature of the molten material decreases as it moves through the mold, the temperature of each tooth during injection molding changes depending on its position from the gate. This temperature distribution causes differences in material shrinkage behavior during the cooling process, which is thought to influence the shape of the tooth flank. However, the effect of changes in manufacturing conditions during injection molding on the tooth flank was not investigated in this study. Future investigation is necessary to evaluate the effect of temperature distribution in the material during injection molding.

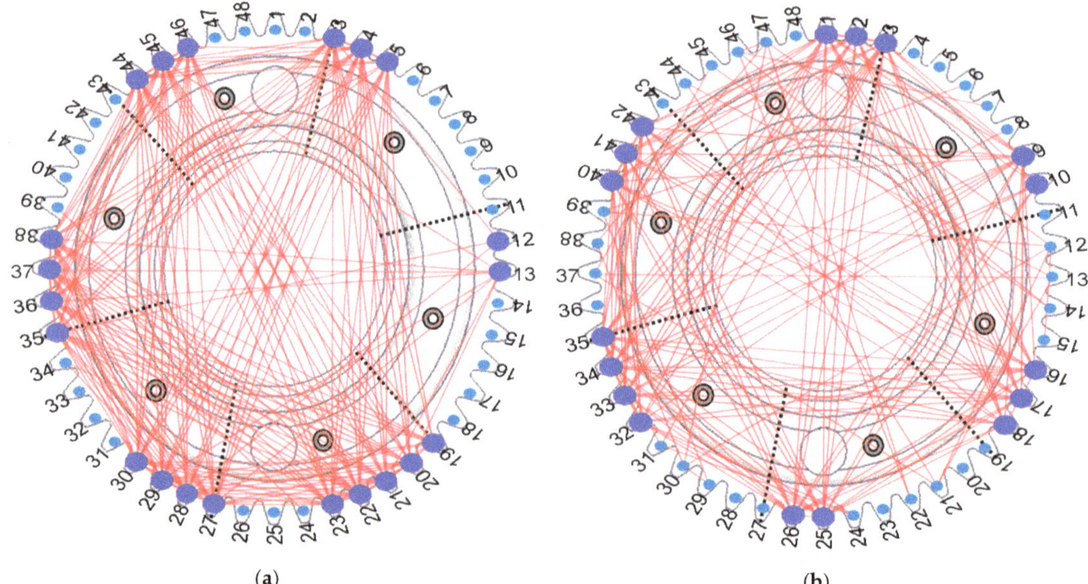

**Figure 14.** Network images of correlation coefficient range of $0.75 \leq r_{j,k} \leq 1.0$ plotted on gear illustration where the six grey circular rings represent gate marks, the six dotted lines represent weld lines, the two white circles represent pin holes: (**a**) Left helix deviation; (**b**) Right helix deviation.

With that being said, from the analysis results obtained from Sections 3.3 and 3.4, the occurrence of uniform and periodic patterns with six high correlation coefficient regions and six low correlation coefficient regions in the pixel plot, and the formation of a network that exhibits a hexagon shape with six concentrated link spots were deeply influenced by the position of the injection molding gate as shown in Figure 14.

## 5. Conclusions

In this study, an analytical methodology has been adopted to assess the relationship between the helix deviations of each tooth in plastic gears manufactured via injection molding [14,15]. The correlation coefficient is computed from the helix deviation curves of one tooth and all the teeth after the elimination of the DC and slope component as a preprocessing measure, and the network was visualized by generating a pixel plot to study the relationship between each gear teeth. Network images were then generated to further dissect the relationship between each gear tooth across varying correlation coefficient ranges. In summary of the analysis result and discussion, the accomplishments of this study are presented as follows:

1.  The relative relationship between each gear tooth can be understood with the application of visualization of network methods such as pixel plot and network image as proposed in this paper.
2.  The helix deviation network of an injection-molded plastic gear is deeply influenced by the position of the injection molding gate of the mold.

The result of this study is vital to acknowledge the influence of the position of the injection molding gate on the tooth helix deviation network of injection-molded gear. he visualization of the helix deviation network of injection-molded plastic gear serves as feedback to its manufacturing process, and appropriate adjustment of injection molding environment and condition can be made with judgment on the topology of the helix deviation network, which can further improve the accuracy of injection-molded gear. The obtained result is useful in the development of a new indicator for the manufacturing error evaluation method of injection-molded plastic gear, and thus enhances the accuracy of the tooth flank of an injection-molded plastic gear.

**Author Contributions:** Conceptualization, D.I., D.Y., J.C.L. and Y.S.; methodology, D.I., D.Y., J.C.L. and Y.S.; software, D.I., D.Y., J.C.L. and Y.S.; validation, D.I. and J.C.L.; formal analysis, J.C.L. and D.I.; investigation, J.C.L. and D.I.; resources, D.I.; data curation, J.C.L., D.Y. and Y.S.; writing—original draft preparation, J.C.L. and D.I.; writing—review and editing, J.C.L. and D.I.; visualization, J.C.L. and D.I.; supervision, D.I.; project administration, D.I.; funding acquisition, D.I. All authors have read and agreed to the published version of the manuscript.

**Funding:** This research was funded by the Japanese Ministry of Education, Culture, Sports, Science and Technology, Grant-in-Aid for Scientific Research (C), 21K03835.

**Institutional Review Board Statement:** Not applicable.

**Informed Consent Statement:** Not applicable.

**Data Availability Statement:** The data presented in this study are available on request from the corresponding author.

**Acknowledgments:** The authors gratefully acknowledge the support from the Japanese Ministry of Education, Culture, Sports, Science and Technology, Grant-in-Aid for Scientific Research (C), 21K03835.

**Conflicts of Interest:** The authors declare no conflicts of interest.

## References


1. Park, C.I. Noise characteristics and acoustic design sensitivity of the helical gear system in wind turbine load. *J. Mech. Sci. Technol.* **2022**, *36*, 5915–5924. [CrossRef]
2. Lee, J.Y.; Chang, J.H. Vibration and noise characteristics of coaxial magnetic gear according to low-speed rotor structure. *J. Mech. Sci. Technol.* **2017**, *31*, 2723–2728. [CrossRef]
3. Park, C.I. Characteristics of noise generated by axial excitation of helical gears in shaft-bearing-plate system. *J. Mech. Sci. Technol.* **2015**, *29*, 1571–1579. [CrossRef]
4. Mohamad, E.N.; Komori, M.; Matsumura, S.; Ratanasumawong, C.; Yamashita, M.; Nomura, T.; Houjoh, H.; Kubo, A. Effect of variations in tooth flank form among teeth on gear vibration and an sensory evaluation method using potential gear noise. *J. Adv. Mech. Des. Syst. Manuf.* **2010**, *4*, 1166–1181. [CrossRef]
5. Ogawa, Y.; Matsumura, S.; Houjoh, H.; Sato, T.; Umezawa, K. Rotational Vibration of a spur gear pair considering tooth helix deviation (Development of simulator and verification). *JSME Int. J. Ser. C Mech. Syst. Mach. Elem. Manuf.* **2000**, *43*, 423–431. [CrossRef]
6. Ryo, E.; Tadami, K.; Kazutaka, S. Mechanism of Occurrence of Ghost Noise in Helical Gears Resulting from Thread-Type Tool Grinding(Machine Elements, Design and Manufacturing). *Trans. Jpn. Soc. Mech. Eng. Ser. C* **2009**, *75*, 2575–2580.
7. Shigeki, M.; Haruo, H. Vibration diagnosis method of multistage gearing to point out ghost noise source with amplitude fluctuation analysis. *Trans. JSME* **2015**, *81*, 15-00004. (In Japanese)
8. Myungsoo, K.; Daisuke, I.; Tomohiro, T.; Junichi, H.; Morimasa, N.; Ichiro, M. New measuring method for single pitch deviation of hobbed gears (Using measured helix form data conditioned with Gaussian filter). *Trans. JSME* **2016**, *82*, 16-00128. (In Japanese)
9. *ISO 1328-1:2013*; Cylindrical Gears—ISO System of Flank Tolerance Classification—Part 1: Definitions and Allowable Values of Deviations Relevant to Flanks of Gear Teeth. ISO: Geneva, Switzerland, 2013; pp. 1–50.
10. *JIS B 1702-1*; Cylindrical Gear-Accuracy Class—Part 1: Definition and Tolerance of Errors on Gear Tooth Flanks. Japanese Industrial Standard: Tokyo, Japan, 2016; pp. 1–20.
11. Barabási, A.-L. Network Science. Chapter 2. Available online: http://networksciencebook.com (accessed on 6 April 2023).
12. Ionnis, P.; Qin, Z. Optimization of injection molding design. Part I: Gate location optimization. *Polym. Eng. Sci.* **1990**, *30*, 873–882.
13. Kawasaki, R.K.; Ikeda, Y. Network analysis of attitudes towards immigrants in Asia. *Appl. Netw. Sci.* **2020**, *5*, 85. [CrossRef]
14. Iba, D.; Noda, H.; Inoue, H.; Kim, M.; Moriwaki, I. Visualization of phase differences between tooth helix deviations using graph theory. *Forsch. Ingenieurwesen* **2019**, *83*, 529–535. [CrossRef]
15. Iba, D.; Inoue, H.; Noda, H.; Kim, M.; Moriwaki, I. Networks of tooth helix deviations of ground and super-finished gears—Phase edges and intensity vertices. *J. Mech. Sci. Technol.* **2019**, *33*, 5689–5697. [CrossRef]

16. Senthilvelan, S.; Gnanamoorthy, R. Fiber Reinforcement in Injection Molded Nylon 6/6 Spur Gears. *Appl. Compos. Mater.* **2006**, *13*, 237–248. [CrossRef]
17. Solanki, B.S.; Singh, H.; Sheorey, T. Study on Gate Location and Gate Number for Manufacturability of Polymer Gears. In *Recent Advances in Mechanical Engineering*; Springer: Singapore, 2021; pp. 71–81.

*Article*

# Integration of Sensors for Enhanced Condition Monitoring in Polymer Gears: A Comparative Study of Acceleration and Temperature Sensors

Sascha Hasenoehrl, Julian Peters and Sven Matthiesen *

Institute of Product Engineering, Karlsruhe Institute of Technology (KIT), 76131 Karlsruhe, Germany; sascha.hasenoehrl@kit.edu (S.H.); julian.peters@kit.edu (J.P.)
* Correspondence: sven.matthiesen@kit.edu; Tel.: +49-721-608-47156

**Abstract:** As an integral part of a machine, gears are subject to wear, which is influenced by a number of factors. For polymer gears in particular, the uncertainties due to wear are high. These uncertainties outweigh the advantages of polymer gears, such as lower inertia. Improved condition monitoring, for example, with better data acquisition, could reduce these uncertainties and is therefore of great interest. This study addresses the challenges of condition monitoring in polymer gears by investigating the integration of sensors directly onto the gears for improved sensitivity. A compact sensor module mounted on a polymer gear is presented to demonstrate the benefits of integrated sensors. The research compares the effectiveness of integrated acceleration and temperature sensors with state of the art external methods. The results show that the in situ sensor module (ISM) provides reliable measurements for condition monitoring with integrated sensors. A comparative analysis with methods based on the current state of research highlights the increased sensitivity of condition monitoring based on the ISM acceleration sensors compared to traditional bearing block sensors. This increased sensitivity shows a clear advantage of integrated sensors over established methods. The temperature curve of the integrated sensors is sensitive to abrasive wear and gear failure, indicating the wider potential of integrated temperature sensors. In conclusion, this research lays the foundation for advanced condition monitoring using integrated sensors in polymer gears. The knowledge gained contributes to optimising gear applications, promoting cost-effectiveness and aligning with the principles of the Internet of Things and Industry 4.0.

**Keywords:** in situ; condition; vibration; analysis; smart gear; thermal behavior; damage

**Citation:** Hasenoehrl, S.; Peters, J.; Matthiesen, S. Integration of Sensors for Enhanced Condition Monitoring in Polymer Gears: A Comparative Study of Acceleration and Temperature Sensors. *Appl. Sci.* **2024**, *14*, 2240. https://doi.org/10.3390/app14062240

Academic Editors: Franco Concli, Aleksandar Miltenovic and Stefan Schumann

Received: 29 January 2024
Revised: 28 February 2024
Accepted: 4 March 2024
Published: 7 March 2024

## 1. Introduction

Gears are one of the most common used machine elements. The wear of gears is highly dependent on the loads, speed and environment it experiences during its life cycle [1]. The geometric changes in the gears due to wear result in a transmission error, which in turn leads to additional transmission losses and noise [2–4]. If the machine is not properly maintained, this wear can cause gear failure, a loss of function and further damage to the machine [3]. Consequently, wear leads to high costs and long periods of downtime. Therefore, it is more economic to monitor gear wear and detect faults early. In addition, polymer gears are highly susceptible to wear, but they have great potential due to their cost-effectiveness and low mass, resulting in low energy requirements in applications with high acceleration change [5]. Better monitoring of the wear of polymer gears could lead to their wider use, resulting in more cost-effective and environmentally friendly gear applications. To achieve better monitoring, research is needed into more sensitive condition monitoring that can reliably detect mild wear, indicate the degree of wear or even predict the remaining useful life [3,6]. Therefore, two points can be addressed: improve the data acquisition process or improve the analysis methods to resolve the actual wear. This work focuses on the first aspect by exploring integrated sensors in gears, which could enable more sensitive condition monitoring through new sensor concepts and higher signal quality. In addition, the use of more sensor-integrated machine elements enables better monitoring of machines in general. General machine monitoring can lead to optimisations of various kinds, such as more efficient production lines. Those connected sensors are of great interest in the context of the Internet of Things and Industry 4.0.

### 1.1. Sensors

The current state-of-the-art method for capturing the data necessary for gear condition monitoring is by using acceleration sensors on the bearing blocks or casing [2,7,8]. The mounting of the sensors means that vibrations are not measured at their source. These measurements are also called ex situ, as opposed to in situ, which means at the source. As a result, the vibration signal contains vibrations from multiple moving elements along the signal path. Furthermore, the longer the signal path, the more dampened the signal gets [2,9,10]. Consequently, the signal quality can be poor, making it difficult to detect mild wear. Chen et al. also emphasized the necessity of monitoring mild wear [6].

However, the approach of integrating sensors into machine elements of the transmission is not new, as several studies show [7,11,12]. In the following, a few such studies are introduced. The design is similar for all of them. The sensors are placed as close as possible to the source of vibration. The sensors are scanned by a central processing unit which temporarily stores or wirelessly transmits the data. Additionally, a power supply module is required. It is not always the case that all of these necessary components are integrated. Smith et al. [11], Lewicki et al. [7] and Peters et al. [12] all presented accelerometers integrated into a gear or a hollow shaft. They showed that it was possible to detect faults like spalling, tooth fracture or heavy wear with in situ measured data. Although Martin et al. suggested an additional benefit from shortening the signal path [9], the benefit could not be demonstrated. This lack of demonstration could be due to poor sensor data as noted by Smith et al. and Peters et al. Lewicki et al. integrated the sensor into a hollow shaft, resulting in a longer signal path than the others. These factors may explain why the quality of condition monitoring was not as good as expected. Another problem is the space required by the solutions presented. Due to the design chosen, the sensors were installed separately from the microcontroller. This design required a second module on the shaft, which takes up extra space. For full integration on the gearwheel, the sensor and the necessary accessories such as microcontroller and data transfer unit need to be even smaller. There are a number of publications dealing with the full integration of sensors into the gearwheel [13,14].

Besides vibration signals, temperature is one of the common quantities which is used for condition monitoring in general [15]. Elforjani et al. used an integrated sensor that measured the temperature of the oil bath. It was shown that it is possible to detect tooth breakage due to higher temperatures [15]. This method is only suitable for measuring temperature where oil bath lubrication is used. The method is not suitable for other types of lubrication such as grease or dry running. In addition, it is not possible to differentiate the fault between several gears in a gearbox. Resendiz-Ochoa et al. conducted a study to classify the abrasive wear of steel gears based on gearbox temperature [16]. Temperature was measured using an infrared camera, and the classification was based solely on features extracted from the camera images. The study did not measure the temperature of individual gears [16]. To overcome the challenge of measuring temperature at individual gears, it would be necessary to integrate the temperature sensor into the gear.

In conclusion, the condition monitoring of gearboxes based on integrated sensors, namely, acceleration sensors, is part of ongoing research. Other sensors with potential, such as temperature sensors, have received less attention. In addition, the benefits of these integrated acceleration sensors have not yet been experimentally proven, although the benefit of integrated sensors, which would be crucial for monitoring mild wear, has been shown in theory.

### 1.2. Polymer Gears

By integrating sensors, it could be possible to monitor even small changes in gears due to wear. This monitoring is even more important with the increasing use of polymer gears. One of the commonly used polymer materials for gears is polyoxymethylene (POM). POM gears have many advantages over steel gears: low cost, low weight and the ability to run without lubrication [5,17]. However, they have a higher wear rate, which depends on the surface temperature of the gear flank [18,19]. For the reliable operation of POM gears over long lifetimes and potentially critical operating conditions, condition monitoring could help to assess gear performance and predict potential failures. For the overall wear process of POM gears, three phases were found. First, there is a running-in phase, then a linear wear phase and finally a rapid wear phase [18,19]. At higher loads, the first two phases become shorter until they are absent at a certain load. It was shown that POM gears may experience this phenomenon when subjected to loads exceeding 10 Nm at an ambient temperature of approximately 25 °C [18,19]. Apart from abrasive wear, thermal wear is the main wear component for a POM/steel pairing [20]. The wear rate is similar to that of a POM/POM combination [18]. However, the running-in phase tends to be shorter for a POM/steel combination than for a POM/POM combination [21].

In conclusion, POM gears can have an advantage over steel gears and the wear mechanisms can be similar. However, due to a higher wear rate and more complex interdependencies, the disadvantages outweigh the advantages and the use of POM gears remains limited.

### 1.3. Vibration Analysis

Vibration analysis is an established method to monitor gear condition [22]. It is based on the acceleration of the gear teeth as they mesh. This process gives the Gear Mesh Frequency (GMF), which is very important for analysis, as well as its harmonics [3,6,23]. The GMF is always the same for two mating gears because both gears are mounted on different shafts and the shaft frequency is dependent on the number of teeth of a gear. The GMF and its harmonics are calculated as follows:

$$GMF_i = i \cdot f_s \cdot n$$
$$i : \text{ order of GMF}$$
$$f_s : \text{ shaft frequency}$$
$$n : \text{ number of teeth}$$

(1)

As the gear teeth change due to heavy wear or damage, the vibration signals change. This change affects especially the amplitude of the $GMF_i$ [4]. Ziaran et al. suggested that harmonics should be considered at least up to the third order [23]. There are common metrics to detect a fault in gears based on this change in the vibration signal. These metrics are called Gear Condition Metrics (GCMs) and differ in their ability to detect severe faults, monitor faults over time and detect different types of wear mechanisms [24,25].

Two of these metrics have been chosen for this study. The GCMs are calculated as described in Table 1. Table 2 lists the components which are part of the regular, differential and residual vibration signal. More information about the metrics can be found in [24,25].

**Table 1.** GCMs used in this study.

| GCM | Formula | | Description |
|---|---|---|---|
| Energy Ratio (ER) | $ER = \dfrac{\sigma(d)}{\sigma(reg)}$ | (2) | $\sigma(d)$: standard deviation of the differential signal of a data sample<br>$\sigma(reg)$: standard deviation of the regular signal of a data sample<br>N: number of data points in a vibration signal |
| NA4* | $NA4^* - \dfrac{\frac{1}{N}\sum_{i=1}^{N}(res_i - \overline{res})^4}{(var(res_{ok}))^2}$ | (3) | res: residual vibration signal<br>$res_{OK}$: residual vibration signal of a data sample of the unworn gear |

**Table 2.** Composites of vibration signals.

| Signal | Composite |
|---|---|
| Regular | Shaft frequency and $GMF_i$ plus sidebands up to fourth harmonic |
| Differential | Normal vibration signal with removed shaft frequency and $GMF_i$ plus sidebands up to fourth harmonic |
| Residual | Normal vibration signal with removed shaft frequency and $GMF_i$ plus first order sidebands |

The effect of different wear patterns on the frequency spectrum was investigated by Ziaran et al. [23] and Amaranth et al. [26]. They found a change in the sidebands of the GMF as the wear progressed from pitting to abrasive. This change shows that the sidebands respond to abrasive wear, which is the main mechanism considered in this work. Cepstrum analysis uses the inverse Fourier transform to find the periodic components of a signal's spectrum, such as sidebands. Formula (4) is used to calculate the real cepstrum, which is used in this paper:

$$C_r = \mathcal{F}^{-1}\{\log(|\mathcal{F}\{f(t)\}|)\}$$
$$C_r : \text{Cepstrum} \tag{4}$$
$$f(t) : \text{vibration signal}$$

Kumar et al. showed that it is possible to detect faults in POM gears using vibration analysis [8]. However, in most cases, it was not possible to distinguish between a healthy gear and pitting on a single tooth using vibration analysis. It should be noted that in some cases, even pitting spread over several teeth indicated an even healthier gear than pitting on a single tooth. The unclear results could be due to the greater damping of non-fibre-reinforced polymers compared to steel [5]. Better monitoring of gears reduces the uncertainty of wear, enabling use in a wider range of applications, the predictive maintenance of machinery and the reuse of gears.

In conclusion, there are several vibration analysis methods for detecting wear with different objectives. It has already been shown that some of these methods can be applied to polymer gears, but the results are difficult to interpret.

### 1.4. Aim of the Contribution

The state of research shows the need to find a way of collecting wear-correlated data in polymer gears, that provides better data quality. Current methods, such as bearing block sensors, may not be sufficient for detecting wear in polymer gears due to potential high signal damping for ductile polymers. Other methods, such as thermal analysis based on external sensors, are unable to detect faults of individual gears within a gearbox. The approach followed in this contribution is to integrate the sensors onto the gear and measure the variables in situ.

Based on the state of research, the research questions that will be investigated in this contribution are as follows:

Can the integration of sensors into a POM gear lead to better condition monitoring of the gear?

1. Can integrated acceleration sensors provide more sensitive condition monitoring of POM gears than external acceleration sensors?
2. Can integrated temperature sensors into the POM gear provide reliable data that correlate with the wear of the gear?

The aim of this paper is to investigate integrated sensors for the wear condition monitoring of a POM gear.

### 2. Materials and Methods

The research questions include the following aspects:

- Comparison of state-of-the-art condition monitoring using external acceleration sensors and sensors integrated onto a polymer gearwheel.

- Comparative measurements using a temperature sensor integrated into a polymer gear and correlating the measurement data with the wear of the gear.

A compact sensor module is presented in this paper. This module is equipped with acceleration and temperature sensors and is mounted on a polymer gear. As part of the objective, the benefit of an integrated acceleration sensor over the state-of-the-art bearing block sensors is investigated through an experimental study. The response of an integrated temperature sensor to wear will also be investigated by comparing sensor data with validation measurements and actual wear. This research can provide the basis for new condition monitoring methods based on multiple integrated sensors.

### 2.1. In Situ Sensor Module

A printed circuit board was developed that incorporates a sensor concept to measure the vibration of the gear in situ. The in situ sensor module (ISM) is shown in Figure 1. It consists of two MEMS acceleration sensors (ADXL1005, Analog Devices, Wilmington, MA, USA) which allow a compact design. More details about the sensors are listed in Table 3. They measure the tangential acceleration on the gear. The sampling rate of these sensors is 16 kHz. In addition, the ISM was fitted with two temperature sensors (TMP1075, Texas Instruments, Dallas, TX, USA). These sensors are connected to copper thermal pads on the underside of the ISM to transfer the heat from the gear. The thermal conductivity of the copper thermal pads is 380 W/m·K. Therefore, the sensors were able to measure the temperature of the gear wheels core surface every 5 min. The ISM is controlled by a microcontroller (STM32WB55RG, ST microelectronics, Geneva, Switzerland). The recorded data is stored in flash memory on the ISM. The storing is due to the sampling rate of the accelerometers, which results in a large amount of measurement data. The microcontroller also allows data to be sent via Bluetooth Low Energy (BLE), which is used to send the data to a host computer for final storage and analysis. The BLE is also used to control the ISM via commands. This control allows data acquisition from the sensors to be started dynamically and with variable measurement durations. In addition, a special command can be used to check the status of the ISM and stop the test bench if an error occurs. All components are mounted directly on the module's circuit board. The circuit board and the microcontroller are powered by two LiPo 120 mAh batteries that are mounted on top of this board using a 3D-printed holder. This design allows the complete ISM to be mounted directly and flush to the gearwheel using screws that pass through the gear, as shown in Figure 1. Without the need for an additional module on the shaft or bearing block, it is more compact than the designs of Smith et al. [11], Lewicki et al. [7] and Peters et al. [12].

(a)

(b)

**Figure 1.** Pictures of the ISM on the POM gear: (a) without batteries (white: microcontroller; blue: flash memory; yellow: accelerometer and direction of measurement; red: temperature sensor); (b) with the battery holder and batteries.

**Table 3.** Properties of the sensors.

| | In Situ Sensor Module (ISM) | Bearing Block Sensors (BBSs) |
|---|---|---|
| Name | ADXL05 | PCB-356A02 |
| Type | MEMS with analog output | Piezo |
| Measurement range | ±100 g | ±50 g |
| Frequency Range | 23 kHz | 5 kHz |
| Resonance Frequency | 42 kHz | 25 kHz |
| Sensitivity | 14.3 mV/g | 10 mV/g |
| Sensitivity change from 20 °C to 50 °C | ±2% | −10–0% |
| Noise Density | 125 µg/$\sqrt{\text{Hz}}$ | 5 µg/$\sqrt{\text{Hz}}$ |

## 2.2. Test Bench Setup

The ISM was mounted directly onto POM gears (Mädler, Stuttgart, Germany). The characteristics of the gears are given in Table 4. The POM gears were fixed to the shaft with a locking assembly. According to the manufacturer, these gears can run at a continuous temperature of up to 100 °C. A steel gear was selected as the mating gear. This selection is a common material paring for gears [20,27]. The material paring results in lower temperatures on the POM gear, which is desired to control the wear rate [21]. For more information on the gears, see Table 4. No lubrication was added to the setup, which is a common mode of operation for polymer gears [27]. The width of the steel gear is smaller than that of the POM gear. This difference means that not the whole POM gear is engaged. The smaller engagement area helps to accelerate the wear progress. Two synchronous motors were selected for the drive and load (CMP71L, SEW-Eurodrive, Bruchsal, Germany). Two piezo sensors (PCB-356A02, PCB Piezotronics, Depew, NY, USA) were mounted on the bearing blocks as a reference measuring system. These sensors will be referred to as Bearing Block Sensors (BBSs). Further information on the piezo sensor is given in Table 3. The test bench configuration is shown in Figure 2. An ADwin Pro II with a T11 CPU was used to record the data and control the test procedure described in Section 2.3.

**Table 4.** Properties of the gears.

| | POM Gear | Steel Gear |
|---|---|---|
| Material | POM | steel |
| Fabrication | milled | milled |
| Module | 1.5 | 1.5 |
| Number of teeth | 55 | 65 |
| Width | 17 mm | 10 mm |
| Diameter | 82.5 mm | 97.5 mm |
| Weight | 165 g | 742 g |
| Thermal conductivity | 0.31 W/m·K | 50.2 W/m·K |

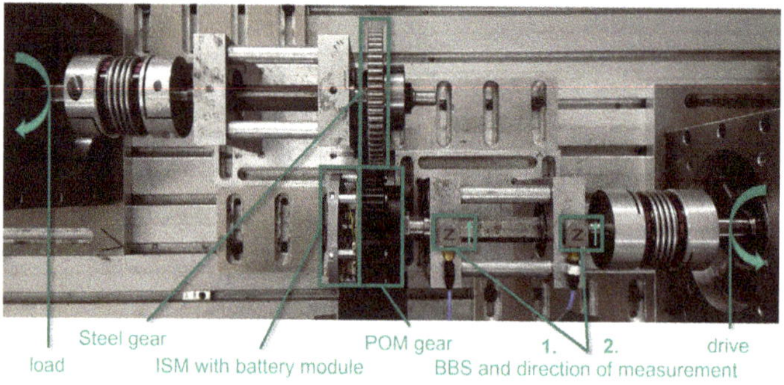


**Figure 2.** Configuration of the test bench.

*2.3. Test Procedure*

For the study in this paper, two POM gears were fitted with an ISM and run alternately on the test bench. In the following, these are called gear 1 and gear 2. The BBSs were used to compare the in situ data with the conventional method of recording vibration data. The sampling rate of the piezo sensors was 10 kHz. The drive ran at a constant 1000 rpm, which is a shaft frequency of 16.7 Hz, and the load was a constant 10 Nm. An overview of the main frequencies is given in Table 5. Both the ISM and the BBSs collected data synchronously for 1 min every 20 min. The test bench was stopped after running for two hours to measure the actual wear and recharge the batteries. These two hours are referred to as two-hour sequences in the following. They consist of $1.2 \times 10^5$ cycles of the polymer gear. After recharging, the gears were remounted and the next two-hour sequence began. Figure 3 displays the flow of the test procedure. Additionally, a 1 min acceleration sample was taken at the beginning of the tests to provide a comparison with the acceleration signal of a completely healthy gear.

**Table 5.** Shaft frequency and $GMF_i$ up to the fourth order.

| Shaft Frequency | $GMF_1$ | $GMF_2$ | $GMF_3$ | $GMF_4$ |
|---|---|---|---|---|
| 16.7 Hz | 916.7 Hz | 1833.3 Hz | 2750 Hz | 3666.7 Hz |

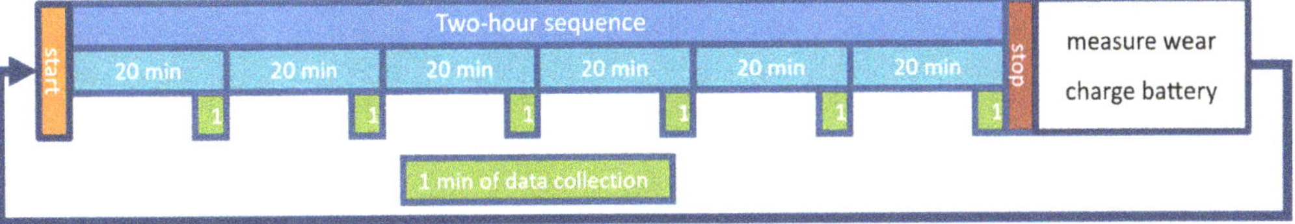


**Figure 3.** Flow of the test procedure.

Polymer gears have a risk of melting at elevated temperatures. The melting would lead to a very fast and abrupt failure. To ensure progressive wear, the temperature had to be below the melting temperature. To confirm that this temperature was not reached in the study, the surface temperature of the loaded flanks of gear 1 was measured. The measurement was also performed to validate the measurements of the integrated temperature sensors on the ISM. The temperature was measured for the second two-hour sequence beginning from 2 h running time. During this period, the temperature was measured every 5 min until a steady-state tooth surface temperature was reached at 3:40 h. The temperature comparison measurements were made using a handheld infrared camera (FLIR E6, FLIR Systems Inc., Wilsonville, OR, USA).

Because of the complex relationship between the vibration signal and wear, it was also necessary to measure the actual wear. For this study, the normed wear was defined as the difference between the measured value and the initial value. To quantify it, the mass loss and the loss of Lateral Surface Area (LSA) was measured. Therefore, the weight of the POM gears was measured using a scale with a precision of $\pm 0.0004$ g (PCE-ABI 220, PCE Deutschland GmbH, Meschede, Germany). The ISM and battery holder were not removed as their weight remained the same. Weighting is a common method for measuring wear [12,28]. The LSA was measured using a digital microscope (VHX2000, Keyence, Osaka, Japan). A similar method was used by Mao et al. who measured the tooth width [29]. The LSA was measured for two different teeth on each POM gear.

*2.4. Data Processing*

Prior to vibration analysis, the in situ data were down-sampled from 16 kHz to 10 kHz. The down-sampling was performed to improve the comparability of the in situ data with the ex-situ data. The aim was to obtain the GMF up to the fourth order, which corresponds to 3.7 kHz. Therefore, 10 kHz is sufficient to satisfy Shannon's sampling theorem [30]. All data were then filtered with a sixth-order Butterworth bandpass. As the fourth-order GMF should be preserved, 4 kHz was used as the upper bound. The lower bound was set to 5 Hz. For the BBS, the tangential measurement direction shown in Figure 2 was used for comparison. The maximum amplitude of each two-hour sequence vibration signal, spectrum, cepstrum and GMCs were then calculated. From Section 2.3, the NA4* metric has an $N$ of 600,000 and the residual of the first 1 min sample was assumed to be $r_{OK}$. The average of the two MEMS sensors on the ISMs and the two BBSs was taken. The results were then compared. The normed wear was calculated by dividing the mass and LSA loss by their respective maximum values. The temperature measurements were used to calculate the static difference temperature by subtracting the ambient temperature from the gear maximum temperature. To determine the increased sensitivity, the quotient of the GCM of the BBS and the ISM was calculated.

## 3. Results

Gear 1 ran for $15.6 \times 10^5$ cycles without a total loss of function. However, the shaft-hub connection of gear 2 showed server damage after $6.8 \times 10^5$ cycles of running, so data analysis was only possible up to that point. The results of this data analysis are presented in the following section, starting with wear, then temperature and finally vibration.

### 3.1. Wear

A picture of the melted inner radius of gear 2 is shown in Figure 4. Figure 5a,b are digital microscope images of gear 1. They show the shrunken Lateral Surface Area (LSA) of the new gear 1 against the worn gear 1. The left side of the LSA in Figure 5b shows that not only is the tooth getting smaller as indicated by the measured area, but the material is also being displaced. Weight measurements taken to monitor the wear of the POM gears are shown in Figure 6a. The weight of gear 1 is higher than that of gear 2 due to a different design of the battery holder. Over the course of the study, the weight of both gears decreases in all measurements. Initially, the decrease is slightly steeper for gear 2 than for gear 1. The LSA shown in Figure 6b decreases overall with a very similar slope for gear 1 and gear 2. There is also an increase from time to time; this phenomenon will be discussed in Section 4.1. Overall, the LSA of tooth 2 of gear 1 decreases slightly faster than the LSA of tooth 1. As can be seen from the approximation of the normed wear in Figure 6c, the loss of weight and LSA of the POM gear has the same linear trend.

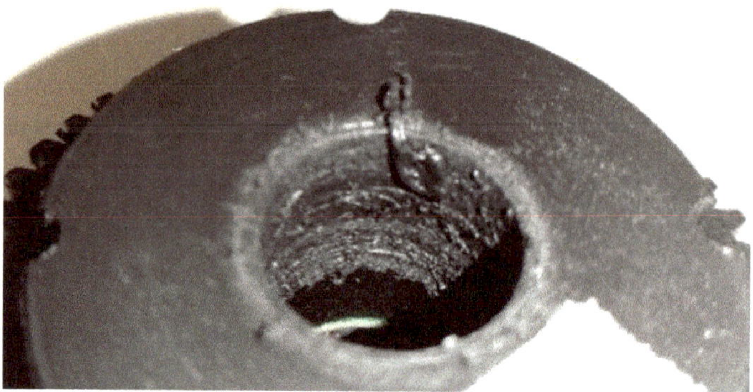

**Figure 4.** Molten shaft hub connection of gear 2 after failure.

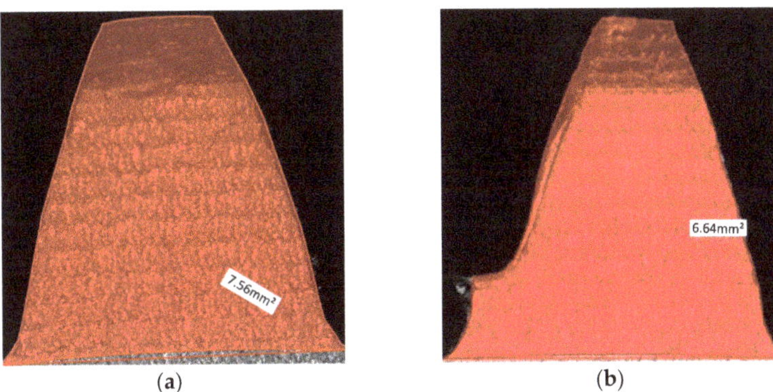


**Figure 5.** LSA 1 of gear 1: (**a**) after 0 cycles; (**b**) after $15.6 \times 10^5$ cycles.

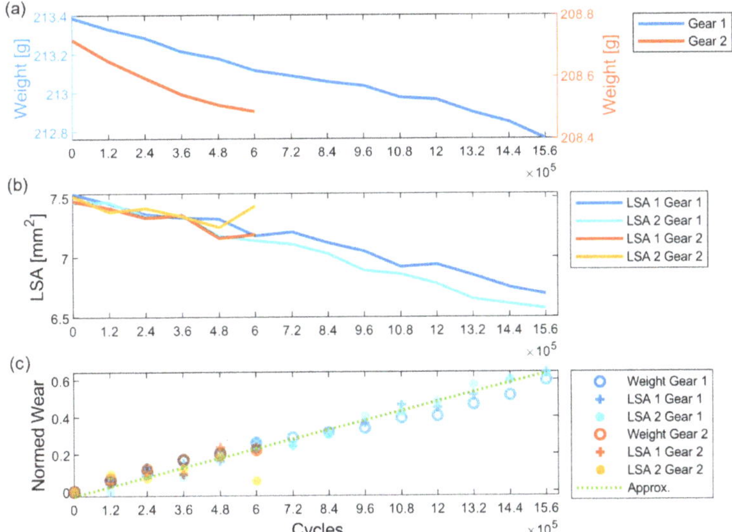

**Figure 6.** Wear of the POM gears: (**a**) weight of the gear plus ISM and battery holder; (**b**) LSA; (**c**) normed wear.

## 3.2. Temperature

Figure 7 shows the temperature over a two-hour sequence. The sequences all look similar, apart from the failure sequence. The temperature followed a bounded growth. Both the reference measurement of the infrared camera and that of gear 1 from 2 h to 3:40 h start at roughly the same room temperature of 22 °C and increase over the sequence. The bound is at about 41 °C and was reached at about 1:20 h into the two-hour sequence. The progression of the temperature curve measured using the ISM at the same time follows the progression of the temperature measured using the infrared camera. This progression also counts for the second temperature curve of gear 1 shown in Figure 7. For gear 2, the temperature measured using the ISM is shown for the two-hour sequence in which the shaft–hub connection failed. The temperature rises more steeply from the beginning, but still follows a bounded growth until 1 h. However, after 1 h of operation, the temperature starts to rise steeply. It reaches a peak at the failure time of 11:20 h of running.

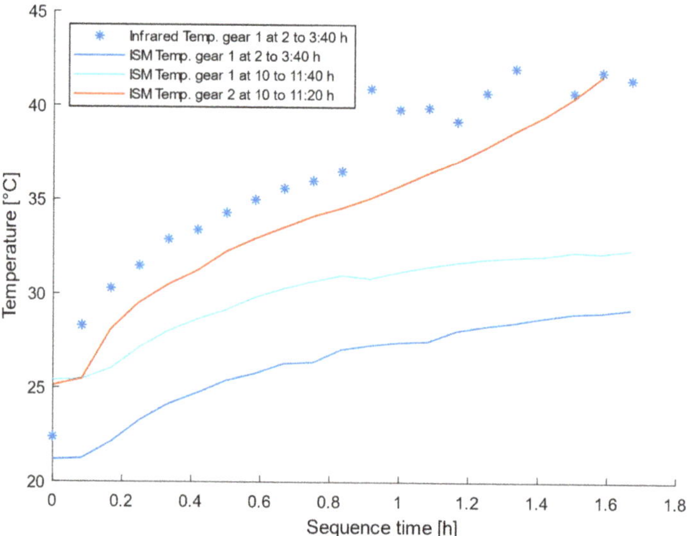

**Figure 7.** Temperature of gear 1 between 2 and 3:40 h running time and gear 1 and 2 between 10 and 11:40 and 11:20 h respectively.

Figure 8 shows the maximum and static difference temperature of each two-hour sequence measured using the ISM. At the beginning, the maximum temperature of both gears increases. This increase applies to both the absolute and the static difference temperature. For the maximum temperature, the increase is about 17%. The static difference temperature increases by about 100%. Then, the static difference temperature decreases by about 50%. The static difference temperature of gear 1 then remains almost constant. In contrast, the shaft hub connection of gear 2 fails and the temperature rises to 42 °C.

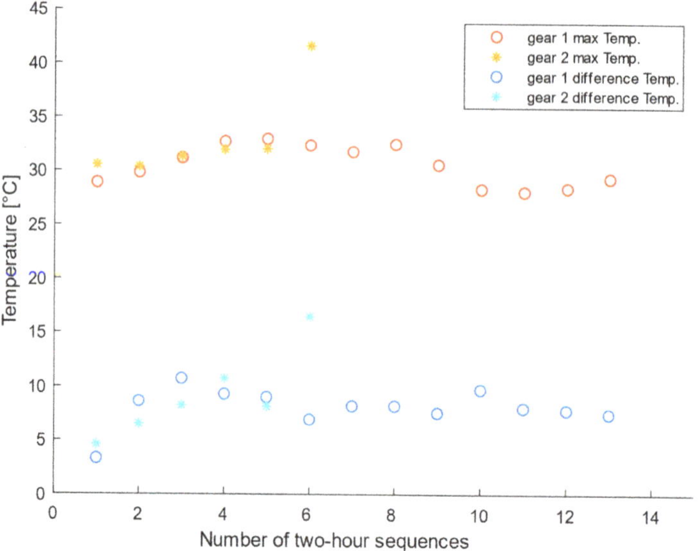

**Figure 8.** Maximum and static difference temperature measured using the ISM of gear 1 and gear 2 over every two-hour sequence.

*3.3. Acceleration Signal Properties*

Figure 9 shows the effect of wear on the acceleration spectrum. The shaft frequency is clearly visible in Figure 9a; $GMF_1$ and $GMF_2$ are visible in Figure 9a,b and all at the frequencies shown in Table 5. The shaft frequency dominates the spectrum of the ISM. However, with BBS, the shaft frequency is low. $GMF_1$ is the dominant frequency. For both sensors, $GMF_2$ is still visible. The amplitude of the higher $GMF_i$ is too low compared to the rest of the spectrum to be visible. As wear progresses, the lower frequencies under 2 kHz increase the most. $GMF_1$ and its sidebands almost double after $2 \times 10^5$ cycles for the ISM. The sidebands reach almost half the intensity of the GMF. For the BBS, the increase of the $GMF_1$ is even greater. However, the sidebands are less than half of the $GMF_1$. Therefore, the sidebands relative to the $GMF_1$ of the BBS are smaller than those of the ISM. In addition, two harmonics increase between the shaft frequency and the $GMF_1$ at frequencies of 366.7 Hz and 550 Hz. The further increase up to $15.6 \times 10^5$ cycles of the $GMF_i$ and their sidebands is very small. Instead, the harmonics between the shaft frequency and the $GMF_1$, which first increased, have now shifted from 366.7 Hz and 550 Hz to 200 Hz and 716.7 Hz, respectively. For both sensor types, it appears that the first GMF increases with wear but the second decreases. The amplitude of the measured acceleration by the ISM is generally much lower compared to the amplitude of the BBS. In the case of $GMF_1$ and $GMF_2$, the amplitudes of the ISM are about one tenth of the amplitudes of the two $GMF_i$ for the BBS.

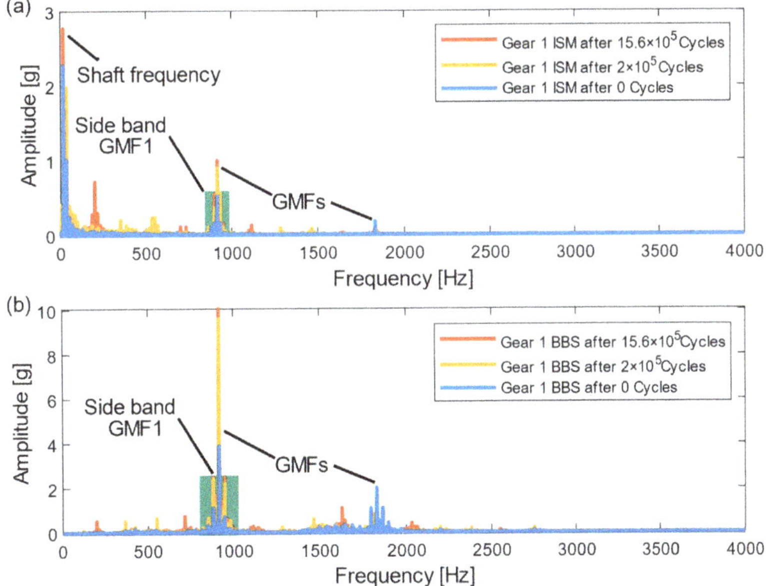


**Figure 9.** Acceleration spectrum of gear 1: (**a**) ISM; (**b**) BBS.

An analysis of the amplitude of the $GMF_1$ confirms these findings. Figure 10 shows that the acceleration amplitude at the $GMF_1$ at 916.7 Hz increases for both gears and both sensor types (ISM, BBS). It shows that there are two phases. Up to about $2 \times 10^5$ cycles, the amplitude increases steeply. After that, the amplitude remains approximately constant for the BBS. However, the amplitude of the ISM drops at the start of several two-hour sequences. This drop does not appear in the amplitude of the cepstrum which represents the sidebands. The amplitude of the ISM $GMF_1$ is about a factor of 10 lower than the amplitudes of the BBS $GMF_1$.

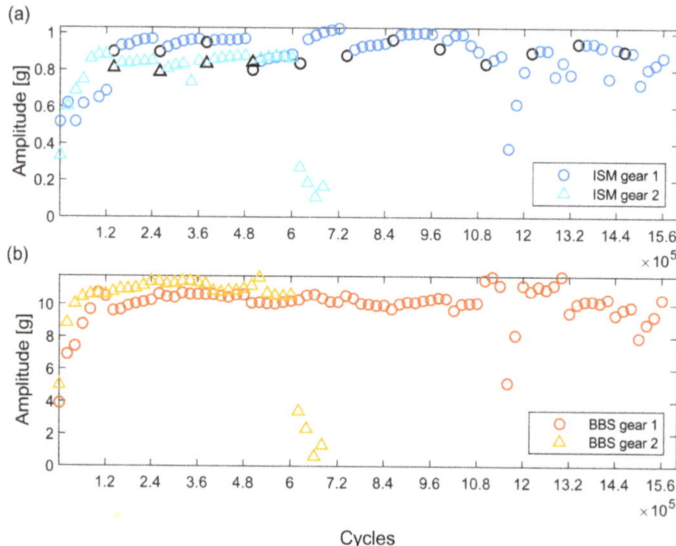

**Figure 10.** Amplitude of the spectrum of the first GMF: (**a**) ISM with the beginning of each two-hour sequence marked in black; (**b**) BBS.

Figure 11 shows that the cepstrum of the $GMF_1$ decreases first. After this first phase, the cepstrum starts to fluctuate and slowly decreases. The amplitudes of the ISM sideband are about a factor of 2 lower than the amplitudes of the BBS sideband. Therefore, the relative height of the sidebands to the GMFs is higher for the ISM than for the BBS. The failure of gear 2 is clearly visible as the amplitude of the $GMF_1$ rapidly decreases and the amplitude of the sideband rises. Figure 12 shows the maximum amplitudes of the vibration signal. It can be observed that the amplitudes measured using the BBS are much higher. Due to an error while saving the data, the ISM data for gear 1 are missing from $0.6 \times 10^5$ to $1 \times 10^5$ cycles in Figures 10a, 11a and 12.

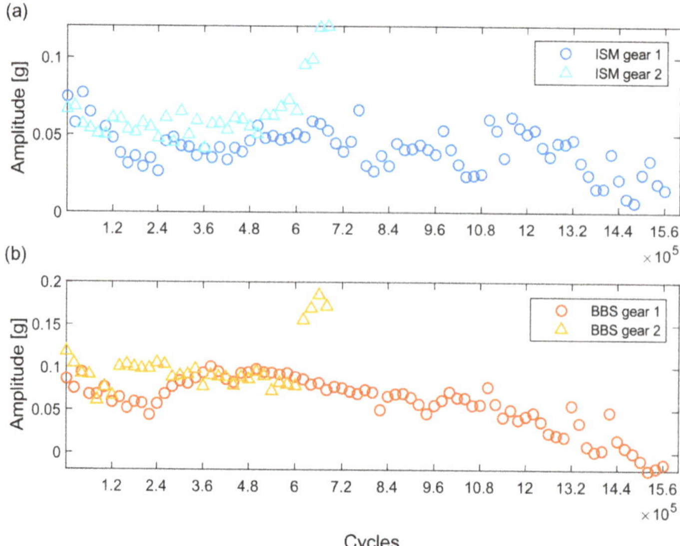

**Figure 11.** Amplitude of the cepstrum of the first GMF: (**a**) ISM; (**b**) BBS.

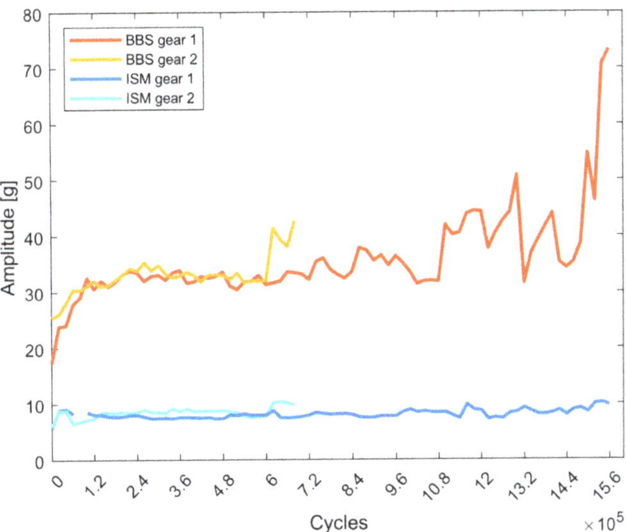

**Figure 12.** Maximum amplitude of each two-hour vibration signal.

### 3.4. Gear Condition Metrics

Figures 13 and 14 show the calculated GCMs normed to their maximum values. There are three phases identified for both GCMs. The first phase is an increase or decrease in the metrics. This phase is followed by a phase of almost constant values. The last phase is an overall increase in the GCMs. In the first phase, the ER acts inversely on the ISM and BBS. The ISM increases and the BBS decreases. An exception to that is the last measurement point in the first phase of gear 2 which is decreasing. The increase of the third phase in the ER starts steeper for the ISM. At this phase, the ER of the ISM is on average 81% higher than that of the BBS.

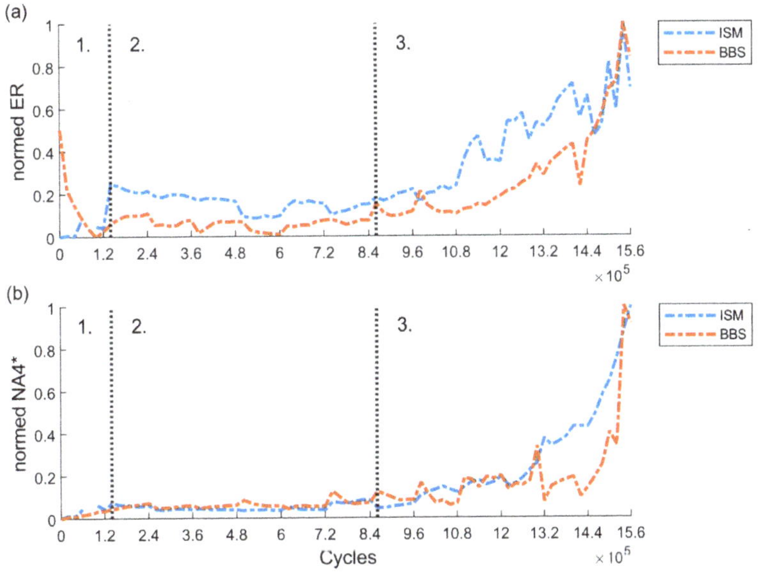

**Figure 13.** GCMs of gear 1 with a separation of the three phases: (**a**) ER; (**b**) NA4*.

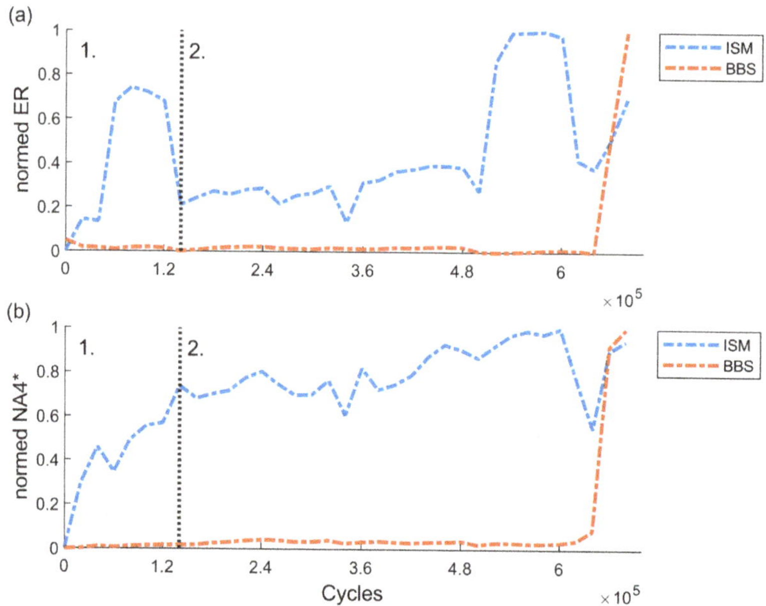

**Figure 14.** GCMs of gear 2 with a separation of the two phases: (**a**) ER; (**b**) NA4*.

In the first phase, the NA4* increases for both sensors. The third phase of the NA4* is similar to that of the ER, with a steeper slope. The NA4* of the BBS is fluctuating much more than the ISM. Due to these fluctuations, the increase of the BBS and ISM seems to be similar in the beginning, but leading to an earlier and more significant increase of the metric for the ISM later into the phase. For the third phase, the NA4* of the ISM is on average about 56% higher than that of the BBS.

In the second and third phases, the GCMs fluctuate strongly at the beginning of a two-hour sequence. These are only visible in the BBS signal. Overall, the NA4* for the ISM is the smoothest. The first two phases appear for both gears. The third phase is not present in the GCMs of gear 2 since it failed before reaching it. The effect of the failure of the second gear shaft-hub connection first causes a steep decrease for the GCMs of the ISM and then an increase. However, the defect is clearly visible by a steep increase in the GCMs of the BBS. The measurement data for cycle $0.6 \times 10^5$ to $1 \times 10^5$ from gear 1 are also missing in the GCMs.

## 4. Discussion

### 4.1. Wear

There was only a slight difference between the two measurements of wear presented in Section 3.1. The slightly steeper decrease in the weight of gear 2 could be due to manufacturing tolerances. As this effect was not found in the LSA measurements, it is more likely to be due to a measurement uncertainty. A possible cause could be that the gears were not properly cleaned prior to measurement and therefore the difference occurred. However, it is noticeable that the Lateral Surface Areas (LSA) of the teeth increased from time to time. This increase was due to the displacement of the material mentioned above. The displacement made it difficult to define the unworn LSA. Otherwise, the wear behaviours of the LSA and the weight were linear and very similar, suggesting that the wear rate was constant.

No running-in phase or constant wear phase was found due to this constant wear rate. The absence is in line with current research, which indicates that the phases are not present beyond 10 Nm at an ambient temperature of approximately between 20 °C and

25 °C [18,19]. The method of measuring wear using the LSA could possibly be improved by using the average tooth width over the worn surface of the teeth. However, as the LSA and weight give similar results and the aim of the study was not to model wear from the measurements, the methods were good enough. Also, the maximum gear temperature recommended by the manufacturer was not exceeded, as shown by the tooth surface temperature measured using the infrared camera. This temperature means that no melting of the gear teeth occurred. A slightly higher wear rate of the LSA of tooth 2 was observed compared to the LSA of tooth 1 of gear 1. Wear is able to spread from teeth with high wear rate to others. This wear rate means that tooth 2 or one of the surrounding teeth of gear 1 could have a small defect.

### 4.2. Temperature Sensor

The temperature of the polymer gear reaches an almost static bound temperature after about 1:20 h into a two-hour sequence. This temperature rise is related to the heat flow that is generated by friction in the contact of the gears, which causes the rise of temperature to the static bound temperature on the flanks of the POM gear of approximately 41 °C. Due to the low thermal conductivity of the gear, a temperature rise also takes place at the measurement point of the ISM, but with a delay and a smaller amplitude. The temperature measured using the ISM on the gear body correlates with the temperature measured using the infrared camera on the tooth flanks. This correlation means that although the temperature values measured using the ISM are lower than the flank temperature, the ISM provides reliable temperature data from which conclusions can be drawn about the tooth flank temperature.

The maximum temperature measured using the ISM was always well below the 100 °C specified by the gear manufacturer. This temperature means that the operating point of the gears was never temperature-critical. The relationship between the static difference temperature curve with its two phases as shown in Figure 8 and the linear wear curve of the POM gear in Figure 6 is not linear. Comparing the evolution of the static difference temperature over time (Figure 8) with the curves of the $GMF_1$ signal characteristics (Figures 9–11), correlations with the temperature curve can be seen. In the first two identified phases of increase in the GMFs and sidebands, the temperature also increases. With the transition to the third phase, the increase in the amplitude of the remaining frequency band, the static differential temperature begins to slowly decrease. This similarity confirms that, similar to Resendiz-Ochoa et al., temperature also has an influence on abrasive wear. Ziaran et al. mapped the different phases of the signal properties to different wear mechanisms [23]. This could mean that the different temperature phases also indicate different wear mechanisms, which causes different friction and therefore different temperatures.

Regarding research question 2 on the integrated temperature sensor, it can be said that the sensor can provide reliable data that correlate with the wear of the POM gear. This correlation is complex and requires further investigation to extract the actual wear condition from the temperature data.

### 4.3. Vibration Analysis

Section 3.3 described how the spectrum of the ISM is dominated by a lower frequency, such as the shaft frequency. In contrast to that, those frequencies are barely visible for the BBS. One aspect of the domination of the ISM by the shaft frequency is due to the rotating coordinate system of the ISM accelerometers within the Earth's gravitational field. The second aspect is the movement of the accelerometer relative to the mesh position of the gear teeth, which causes a change in the signal path of the vibration and therefore a change in the damping of the signal. Both of these movements occur at the shaft frequency, resulting in a high content of the shaft frequency in the ISM spectrum. As the BBS does not move in relation to the earth's gravity or the mesh position of the gears, the content of the shaft frequency in the spectrum is much lower. This phenomenon was also found by Lewicki et al. The signal properties and the GCM do not show the same linear behavior as the wear. This discrepancy shows the complex relationship between the wear and the vibration signal. The results of the vibration analysis can be divided into three phases. These phases are visible in the spectrum shown in Section 3.3 and also in the GCMs. The first phase is accruing because of the changes in the GMF and their sidebands. In the second phase, the changes in the vibration signal are too small to have an impact on the GCMs. The last phase is based on an increase in the amplitudes of frequencies between the $GMF_i$ values. Those are part of the differential and the residual signal, which is why the Energy Ratio (ER) and NA4* are increasing.

The increase in GCMs of the ISM in the third phase started earlier and steeper than that of the BBS. This earlier increase emphasizes that the ISM is more sensitive to changes in the gear than the BBS. In the third phase, the sensitivity of the ISM to the measured wear was on average 81% higher for the ER and 56% higher for the NA4*. This higher sensitivity could be due to the shorter signal path resulting in a less damped vibration signal. In Section 3.3, it was shown that the amplitudes of the sidebands relative to the $GMF_i$ of the BBS are much smaller than those of the ISM. These lower sidebands are another sign of the strong dampening effect of the signal path and better signal quality of the ISM. In addition, the longer signal path leads to more interference from other vibrations. These would mainly affect the differential signal. As a result, the NA4* of the BBS fluctuates more. This fluctuation makes the interpretation of the data more difficult. As a result, fault detection may be more robust with the GCMs based on the vibration data measured with the ISM.

The observed higher amplitudes measured using the BBS contradict the idea that the ISM is more sensitive. However, the amplitudes measured are so high that they are not very plausible. As the characteristics of the sensors have been chosen to suit the described application and there is experience of these sensors in other applications, it is assumed that the fault is in the setup. With the BBS, this fault could be the amplifier or the supply voltage. With the ISM, this fault could be the layout of the components or the power consumption. Furthermore, the comparison between the ISM and the BBS was made on a relative basis, so the absolute values do not contradict the results. The reason for the higher amplitudes of the BBS could not be found.

As shown, the phases identified in the vibration signals do not correlate to the wear rate. Ziaran et al. mapped the different phases of the signal properties to different wear mechanisms [23]. Different wear mechanisms could also be the reason for the different phases found in this study. A hint gives the maximum temperature at each two-hour sequence. Since it is increasing along the first phase identified in the vibration signal, this could mean that in this phase, a more temperature-dependent wear mechanism is present. Evidence for the presence of this mechanism cannot be provided, because no wear mechanisms were reviewed in this study.

In Section 3.3, it was described that the $GMF_1$ drops at the beginning of most of the two-hour sequences. This drop is about 0.1 g out of about 1 g, or about 10%. The temperature change in the ISM over a two-hour sequence is in the range of 20 °C to 50 °C, which can have a total effect on the sensitivity of the acceleration sensors on the ISM of about 4%. As the PCD material is a good thermal insulator, the temperature range and sensitivity deviation should be even smaller. This effect means that the deviation of the $GMF_1$ at the beginning of each two-hour sequence is not necessarily caused by the change in sensitivity due to temperature, but it may be part of the cause. As described in Section 2.2, the POM gears did not run on their whole tooth surface. Figure 15 shows the resulting half-worn tooth surface. To measure wear, the gears had to be removed after each two-hour sequence. When reinstalled, the position of the gears slightly changed axially. Therefore, the gears ran on a partially unworn tooth surface at the beginning of some two-hour sequences. This change in surface could be another possible reason for the drops in $GMF_1$.

Regarding research question 1 on the integrated acceleration sensors, it can be said that the vibration analysis based on the ISM provides more sensitive condition monitoring than the BBS. The higher sensitivity may allow milder wear to be monitored more reliably, which is the first step towards better monitoring of the wear condition of POM gears.

**Figure 15.** Tooth surface of the worn POM gear 1.

## 5. Conclusions

This study investigated the sensitivity of the ISM for abrasive condition monitoring on POM gears and explored the added benefit of integrated temperature and acceleration sensors.

The analysis of wear patterns revealed a consistent linear trend, indicating a constant rate of wear. It was concluded that there were no distinct running-in or constant wear phases, aligning with current research suggesting the absence of such phases at ambient temperatures and high loads. The integrated temperature sensors demonstrated the ability to reliably measure gear temperature. The temperature analysis showed bounded growth, reaching a static bound temperature after 1:20 h into a two-hour sequence. The correlation between ISM and infrared camera measurements confirmed the reliability of the ISM temperature data. The increase of the temperature, along with the sidebands, suggested a possible temperature-dependent influence on abrasive wear. A vibration analysis using the ISM highlighted its sensitivity to changes in gear conditions. The dominant presence of lower frequencies, such as the shaft frequency, in the ISM spectrum distinguished it from BBS. Three phases were identified in the GCMs. The ISM showed less variation and higher sensitivity in the results compared to the BBS, indicating the increased robustness of gear condition monitoring. This publication also discussed the potential influence of temperature on ISM accelerometers, which should be considered in future studies.

This study addresses two detailed research questions. Firstly, the study confirms that the vibration analysis based on ISM provides more sensitive condition monitoring, allowing a potentially better assessment of POM gear wear conditions compared to BBS. Secondly, the integrated temperature sensor as part of the ISM provides reliable data that correlate with POM gear wear. Temperature variations over time are consistent with wear characteristics, highlighting the potential of the sensor to assess gear condition.

While this study focused on a high rate of abrasive wear, future research should investigate different wear mechanisms, especially tooth root fracture, which is of high interest. Therefore, the potential benefits of new integrated sensors need to be further explored to improve condition monitoring. Additionally, new models for assessing actual wear based on new integrated sensors must be investigated to achieve better wear monitoring and potentially calculate the remaining useful life of a POM gear. To better understand the uncertainties that come with such wear monitoring based on integrated sensor data and a model, a higher sample size should be investigated in the future.

**Author Contributions:** Conceptualization, J.P.; methodology, S.H.; software, S.H.; validation, S.H.; formal analysis, S.H.; investigation, S.H.; resources, S.M.; data curation, S.H.; writing—original draft preparation, S.H.; writing—review and editing, J.P.; visualization, S.H.; supervision, S.M.; project administration, S.M.; funding acquisition, S.M. All authors have read and agreed to the published version of the manuscript.

**Funding:** This research was funded by Dr.-Ing Willy-Höfler-Stiftung. The APC was funded by KIT-Publikationsfonds.

**Institutional Review Board Statement:** Not applicable.

**Informed Consent Statement:** Not applicable.

**Data Availability Statement:** The data that support the findings of this study are available from the corresponding author upon reasonable request.

**Conflicts of Interest:** The authors declare no conflicts of interest.

## Abbreviations

| Term | Abbreviation |
| --- | --- |
| Polyoxymethylene | POM |
| Gear Mesh Frequency | GMF |
| Gear Condition Metrics | GCMs |
| Energy Ratio | ER |
| In situ Sensor Module | ISM |
| Bluetooth Low Energy | BLE |
| Bearing Block Sensors | BBSs |
| Lateral Surface Area | LSA |

## References

1. Eyre, T.S. Wear characteristics of metals. *Tribol. Int.* **1976**, *9*, 203–212. [CrossRef]
2. Smith, J.D. *Gear Noise and Vibration*, 2nd ed.; Review and Expanded; Marcel Dekker: New York, NY, USA; Basel, Switzerland, 2003; ISBN 0-8247-4129-3.
3. Feng, K.; Ji, J.C.; Ni, Q.; Beer, M. A review of vibration-based gear wear monitoring and prediction techniques. *Mech. Syst. Signal Process.* **2023**, *182*, 109605. [CrossRef]
4. Kuang, J.H.; Lin, A.D. The Effect of Tooth Wear on the Vibration Spectrum of a Spur Gear Pair. *J. Vib. Acoust.* **2001**, *123*, 311–317. [CrossRef]
5. Mott, R.L.; Vavrek, E.M.; Wang, J. *Machine Elements in Mechanical Design*, 6th ed.; Pearson Education: New York, NY, USA, 2018; ISBN 978-0-13-444118-4.
6. Chen, X.; Wang, S.; Qiao, B.; Chen, Q. Basic research on machinery fault diagnostics: Past, present, and future trends. *Front. Mech. Eng.* **2018**, *13*, 264–291. [CrossRef]
7. Lewicki, D.; Lambert, N.A.; Wagoner, R.S. *Evaluation of MEMS-Based Wireless Accelerometer Sensors in Detecting Gear Tooth Faults in Helicopter Transmissions*; NASA: Washington, DC, USA, 2015.
8. Kumar, A.; Parey, A.; Kankar, P.K. Vibration based fault detection of polymer gear. *Mater. Today Proc.* **2021**, *44*, 2116–2120. [CrossRef]
9. Martin, G.; Vogel, S.; Schirra, T.; Vorwerk-Handing, G.; Kirchner, E. Methodical Evaluation of Sensor Positions for Condition Monitoring of Gears. In *NordDesign 2018 Proceedings of the NordDesign 2018, Linköping, Sweden, 14–17 August 2018*; Ekströmer, P., Schütte, S., Ölvander, J., Eds.; The Design Society: Glasgow, UK, 2018; ISBN 978-91-7685-185-2.
10. Chin, Z.Y.; Borghesani, P.; Smith, W.A.; Randall, R.B.; Peng, Z. Monitoring gear wear with transmission error. *Wear* **2023**, *523*, 204803. [CrossRef]
11. Smith, W.; Deshpande, L.; Randall, R.; Li, H. Gear diagnostics in a planetary gearbox: A study using internal and external vibration signals. *Int J Cond. Monit.* **2013**, *3*, 36–41. [CrossRef]
12. Peters, J.; Ott, L.; Dörr, M.; Gwosch, T.; Matthiesen, S. Sensor-integrating gears: Wear detection by in-situ MEMS acceleration sensors. *Forsch. Ingenieurwes* **2021**, *86*, 421–432. [CrossRef]
13. Binder, M.; Stapff, V.; Heinig, A.; Schmitt, M.; Seidel, C.; Reinhart, G. Additive manufacturing of a passive, sensor-monitored 16MnCr5 steel gear incorporating a wireless signal transmission system. *Procedia CIRP* **2022**, *107*, 505–510. [CrossRef]
14. Bonaiti, L.; Knoll, E.; Otto, M.; Gorla, C.; Stahl, K. The Effect of Sensor Integration on the Load Carrying Capacity of Gears. *Machines* **2022**, *10*, 888. [CrossRef]
15. Elforjani, B.; Xu, Y.; Brethee, K.; Wu, Z.; Gu, F.; Ball, A. Monitoring gearbox using a wireless temperature node powered by thermal energy harvesting module. In Proceedings of the 23rd International Conference on Automation and Computing (ICAC), Huddersfield, UK, 7–8 September 2017; pp. 1–6. [CrossRef]
16. Resendiz-Ochoa, E.; Saucedo-Dorantes, J.J.; Benitez-Rangel, J.P.; Osornio-Rios, R.A.; Morales-Hernandez, L.A. Novel Methodology for Condition Monitoring of Gear Wear Using Supervised Learning and Infrared Thermography. *Appl. Sci.* **2020**, *10*, 506. [CrossRef]
17. Alharbi, K.A.M. Wear and Mechanical Contact Behavior of Polymer Gears. *J. Tribol.* **2019**, *141*, 011101. [CrossRef]

18. Hooke, C.J.; Mao, K.; Walton, D.; Breeds, A.R.; Kukureka, S.N. Measurement and Prediction of the Surface Temperature in Polymer Gears and Its Relationship to Gear Wear. *J. Tribol.* **1993**, *115*, 119–124. [CrossRef]
19. Mao, K.; Li, W.; Hooke, C.J.; Walton, D. Polymer gear surface thermal wear and its performance prediction. *Tribol. Int.* **2010**, *43*, 433–439. [CrossRef]
20. Matkovič, S.; Kalin, M. Effects of slide-to-roll ratio and temperature on the tribological behaviour in polymer-steel contacts and a comparison with the performance of real-scale gears. *Wear* **2021**, *477*, 203789. [CrossRef]
21. Pogačnik, A.; Kalin, M. Parameters influencing the running-in and long-term tribological behaviour of polyamide (PA) against polyacetal (POM) and steel. *Wear* **2012**, *290–291*, 140–148. [CrossRef]
22. Hameed, Z.; Hong, Y.S.; Cho, Y.M.; Ahn, S.H.; Song, C.K. Condition monitoring and fault detection of wind turbines and related algorithms: A review. *Renew. Sustain. Energy Rev.* **2009**, *13*, 1–39. [CrossRef]
23. Ziaran, S.; Darula, R. Determination of the State of Wear of High Contact Ratio Gear Sets by Means of Spectrum and Cepstrum Analysis. *J. Vib. Acoust.* **2013**, *135*. [CrossRef]
24. Lebold, M.; McClintic, K.; Campbell, R.; Byington, C.; Maynard, K. Review of Vibration Analysis Methods for Gearbox Diagnostics and Prognostics. In Proceedings of the 54th Meeting of the Society for Machinery Failure Prevention Technology, Virginia Beach, VA, USA, 1–4 May 2000.
25. Večeř, P.; Kreidl, M.; Šmíd, R. Condition Indicators for Gearbox Condition Monitoring Systems. *Acta Polytech.* **2005**, *45*, 35–43. [CrossRef] [PubMed]
26. Amarnath, M.; Praveen Krishna, I.R.; Krishnamurthy, R. Experimental Investigations to Study the Effectiveness of Cepstral Features to Detect Surface Fatigue Wear Development in a FZG Spur Geared System Subjected to Accelerated Tests. *Arch. Acoust.* **2023**, *46*, 479–489. [CrossRef]
27. Evans, S.M.; Keogh, P.S. Efficiency and running temperature of a polymer–steel spur gear pair from slip/roll ratio fundamentals. *Tribol. Int.* **2016**, *97*, 379–389. [CrossRef]
28. Hu, C.; Smith, W.A.; Randall, R.B.; Peng, Z. Development of a gear vibration indicator and its application in gear wear monitoring. *Mech. Syst. Signal Process.* **2016**, *76–77*, 319–336. [CrossRef]
29. Mao, K.; Li, W.; Hooke, C.J.; Walton, D. Friction and wear behaviour of acetal and nylon gears. *Wear* **2009**, *267*, 639–645. [CrossRef]
30. Shannon, C.E. Communication in the Presence of Noise. *Proc. IRE* **1949**, *37*, 10–21. [CrossRef]

*applied sciences*

MDPI

*Article*

# A Novel Method for the Analysis and Optimization of End Face Stress in Cycloidal Gears Based on Conformal Mapping

Ning Jiang [1], Liquan Jiang [2], Jie Zhang [3] and Shuting Wang [1,*]

[1] School of Mechanical Science and Engineering, Huazhong University of Science & Technology, 1037 Luoyu Road, Wuhan 430070, China; jiangning@hust.edu.cn
[2] The State Key Laboratory of New Textile Materials and Advanced Processing Technologies, Wuhan Textile University, 1 Yangguang Avenue, Wuhan 430200, China; lqjiang@wtu.edu.cn
[3] School of Energy and Power Engineering, Huazhong University of Science & Technology, 1037 Luoyu Road, Wuhan 430070, China; jiezhang@hust.edu.cn
* Correspondence: wangst@hust.edu.cn

**Featured Application: This research presents a novel method for optimizing stress distribution in gear designs, particularly improving the accuracy of intricate geometries like cycloidal gears.**

**Abstract:** In the realm of precision engineering, the evaluation of stress distributions in complex geometries, such as the end face of gears, is paramount for determining performance characteristics, including the fatigue life of transmission devices. This research introduces a methodology that combines the method of conformal mapping for multiply connected regions with a proposed global scaling factor. This combination facilitates the transformation of stress computations from intricate geometries to a unit circle. By employing this method as a foundation for surrogate models, and by incorporating the Bayesian-enhanced least squares genetic algorithm optimization technique, this study refines the end face stress distribution, particularly considering variations induced by tooth profile modification errors. The results delineate a design parameter domain that is conducive to achieving uniform stress distribution, offering a refined perspective and methodology for the design of cycloidal wheels.

**Keywords:** cycloidal gear; end face stress; conformal mapping; surrogate model; genetic algorithm

**Citation:** Jiang, N.; Jiang, L.; Zhang, J.; Wang, S. A Novel Method for the Analysis and Optimization of End Face Stress in Cycloidal Gears Based on Conformal Mapping. *Appl. Sci.* **2023**, *13*, 11805. https://doi.org/10.3390/app132111805

Academic Editors: Franco Concli, Aleksandar Miltenovic and Stefan Schumann

Received: 5 October 2023
Revised: 24 October 2023
Accepted: 25 October 2023
Published: 28 October 2023

## 1. Introduction

Stress distribution plays a pivotal role in the design of gear end surfaces [1–3], profoundly affecting the gear's performance and lifespan. Two primary determinants shape this distribution: the applied load and the material properties of the gear. Typically, the loads imposed on gears originate from power sources such as engines or motors on the input shaft. These loads give rise to compressive and shear stresses on the gear, culminating in varied stress distributions. Furthermore, the intrinsic attributes of the gear material, including its strength, hardness, and toughness, further delineate this distribution.

Stress distribution in gears plays a pivotal role in determining their longevity, reliability, and susceptibility to fatigue. When gears are subjected to loads, the distribution of stress can significantly influence their fatigue lives and the associated risk of fracture [4–6]. An optimized stress distribution enhances the gear's lifespan and reduces failure risks. Many studies have emphasized the importance of understanding the underlying failure mechanisms in gear systems to enhance their reliability, efficiency, and longevity, with fatigue failure, particularly contact and bending fatigue, being a recurring concern. Guan and Wang utilized FEA to delve into the stress distribution in high-speed helical gear shafts, pinpointing fatigue fracture as the primary failure mode [7]. Similarly, the research by Qian et al. indicated that the maximum stress location correlates with the failure site [8]. FEA also surfaced as a crucial tool in other studies, such as those by Jalaja et al., which inspected aluminum alloy brake reducers for reusable launch vehicles, and by Yang et al., who explored stress sensitivity and fatigue life in harmonic gear drives. These analyses underlined the necessity to grasp material attributes and grain orientation to foresee ductile overload or fatigue failures [9]. In a contrasting perspective, Yin et al. highlighted the repercussions of misalignment, such as tooth breakage and pitting due to surface contact fatigue [10]. Meanwhile, Tsai et al. concentrated on structural stress analysis techniques and the prediction of fatigue life in high reduction ratio reducers, juxtaposing FEA-derived outcomes with formula-calculated stress and fatigue life estimates [11]. Furthermore, specific fatigue forms, like rolling contact fatigue in RV reducer crankshafts, have been investigated. Li et al. introduced a 3D elasto-plastic contact model, factoring in hardness gradients and initial residual stress for better fatigue life prediction [12]. Additionally, Feng et al. examined the breakdown of a secondary driving helical gear in electric vehicles, linking the failure to elevated contact stress and impact load caused by subpar material properties [13]. Collectively, these studies underscore the multifaceted aspects of gear failures. Addressing fatigue requires a holistic comprehension of stress distribution, material characteristics, and operational conditions. Consequently, the use of advanced analytical tools like FEA becomes indispensable in designing, diagnosing, and refining gear systems across various applications.

Stress distribution in gears profoundly impacts their transmission efficiencies, closely relating to gear friction, wear, and energy losses [14,15]. Optimizing this distribution can lead to improved transmission and energy conversion efficiencies. A surge of research has focused on the interplay between stress and transmission performance in various reducers, aiming to boost efficiency, reliability, and load capacity through innovative mechanical designs and mathematical modeling. Sun and Han's examination of the China Bearing Reducer (CBR) underscored the significance of the clearance between the cycloid teeth and pin on the contact force and stress [16]. Using finite element methods, Song et al. found that in RV reducers, the eccentric shaft's stiffness is of greater concern than the material strength [17]. Hsieh and team compared traditional and novel designs for two-stage speed reducers, analyzing their implications for stress and motion [18]. Mo et al. showcased the 2K-H internal meshing abnormal cycloidal gear (ACG) planetary reducer's superior transmission properties and reduced stress levels [19], while Hsieh proposed a distinctive transmission design that optimized stress distribution [20]. The analytical models by Li et al. and Kim et al. took into account factors like elastic- and heat-induced deformations in cycloid and planetary gear speed reducers [21]. Huang et al.'s focus was on dynamic characteristics, using both static and dynamic finite element analyses to evaluate the stress distribution [22]. Delving into specific applications, Bao et al.'s finite element analysis led to a notable reduction in the contact stress in cycloid gears [23].

Moreover, the acoustic properties of gears are also affected by stress distribution. Fluctuations in stress can alter a gear's vibration and noise levels, implying that a meticulously crafted stress distribution design can diminish these acoustic interferences, thereby improving the gear's acoustic performance [24]. In summation, recognizing and addressing stress distribution in gears, especially on their end surfaces, is vital. A competent design that considers this aspect can not only refine the performance and reliability of the gear but also uplift the acoustic attributes, thereby optimizing the overall functionality of the mechanical system.

Despite the extensive research on gear stress distribution, several gaps persist, which our investigation aims to bridge. Firstly, while most of the prevailing studies lean heavily on FEA to interpret stress and fatigue, the wide dimensional span between gear profile modifications and the actual gear geometry poses computational challenges. Mature FEA software, when applied to these scenarios, is computationally intensive, making efficient analysis and optimization elusive. To counteract this, advanced optimization techniques like the Bayesian-enhanced least squares genetic algorithm optimization method (BLS-GAOM) emerge as imperative tools, offering precise and cost-effective analyses. Secondly, the assumption of idealized conditions in many studies often overlooks real-world intricacies like input crank deformation, pin tooth runout, and backlash angles. These variables, aside from their individual impacts, can also lead to altered contact characteristics, affecting the overall gear performance. Thirdly, there is a discernible focus on tooth surface contact in the literature, sidelining the more complex yet crucial aspect of stress analysis on gear end surfaces with intricate shapes. This oversight potentially masks a significant area of stress accumulation, which might be the precursor to premature gear failures or sub-optimal performance. In summary, while the field has made commendable strides, a holistic, integrated approach that amalgamates advanced computational methods, real-world variables, and a deep dive into lesser-explored gear regions is the next frontier in gear stress distribution research.

This study is methodically structured to offer a systematic understanding of cycloid gear stress analysis and optimization. Initially, this paper lays the groundwork with a quadratic contact surface analysis and force modeling (Section 2). We then delve into an advanced analysis of end face stress in cycloid gears, introducing a conformal mapping technique that is suitable for multiply connected domains and discussing the geometric intricacies of cycloid gear profiles with pores (Sections 3.1 and 3.2). The novel SACGES-CMMD method is presented (Section 3.3), followed by insights into the global stress scaling factor (Section 3.4). Section 4 propounds an innovative optimization approach, integrating Bayesian analysis with a genetic algorithm. Comprehensive results, ranging from the impact of clearance on contact force fluctuations to a comparative analysis of SACGES-CMMD versus conventional FEM, are detailed in Section 5. This study culminates in Section 6, succinctly summarizing the key findings and their broader implications.

## 2. Quadratic Contact Surface Analysis and Force Modeling

Specifically, studying the second-order surface of gear contact is of paramount significance in optimization, enhancing performance, and preventing malfunctions. Initially, this method aids in comprehending the shape variation and load distribution on the gear contact surface in real-time engagement, facilitating the optimization of mechanical properties like enhanced load-bearing capacity, noise reduction, and wear diminution. Furthermore, insights from the second-order surface can drive design optimizations, for instance, reshaping the gear to minimize sliding friction. Comprehending the characteristics of the gear contact surface also paves the way for designing intricate and efficient gear systems, like helical and bevel gears. Ultimately, understanding the gear contact via the second-order surface can illuminate the wear and failure mechanisms, which are critical for proactive fault prevention and gear maintenance.

In analyzing the contact of cycloid and pin gears, a simplified model that equivalently describes the contact as two cylindrical bodies interacting can be envisioned, as illustrated in Figure 1. The contact between the two cylindrical shapes can be characterized by a second-order equation, represented by the general form as Equation (1):

$$ax^2 + by^2 + cz^2 + 2fyz + 2gzx + 2hxy + 2ux + 2vy + 2wz + d = 0 \tag{1}$$

Given that this equation passes through the origin, by substituting $x = 0$, $y = 0$, and $z = 0$ into Equation (1), it can be deduced that $d = 0$. The point of origin is a mutual tangent plane for both cylindrical bodies, characterized by $x = 0$, $y = 0$, $z = 0$, and $\frac{\partial z}{\partial x} = 0$. The partial differentiation of Equation (1), with respect to $x$ yields, is $\mu = 0$.

Similarly, by differentiating with respect to $y$, we deduce that $v = 0$. With these outcomes, Equation (1) can be condensed into its simplified representation, labeled Equation (2).

$$ax^2 + by^2 + cz^2 + 2fyz + 2gzx + 2hxy + 2wz = 0 \tag{2}$$

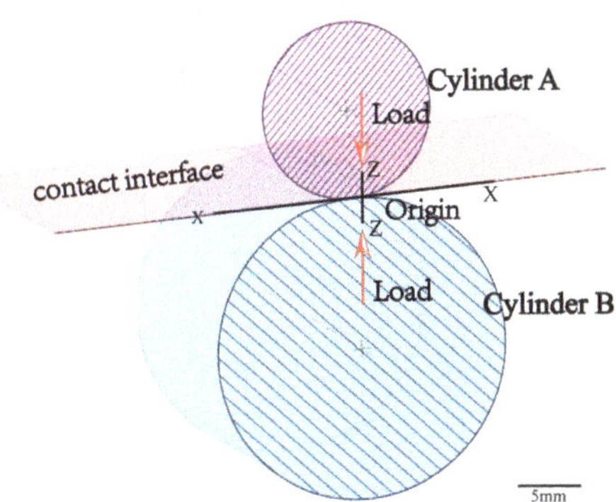



**Figure 1.** Two cylinders meshing.

Having obtained the first-order derivatives of the contact surface, and drawing on Taylor's theorem, it becomes imperative to compute the second-order derivatives to fully specify the equation representing the contact surface. Thus, the second-order partial derivatives of Equation (2) with respect to both $x$ and $y$ need to be calculated, which lead to Equations (3) and (4), respectively.

$$2a + 2c\left(\frac{\partial z}{\partial x}\right)^2 + 2cz\frac{\partial^2 z}{\partial x^2} + 2fy\frac{\partial^2 z}{\partial x^2} + 2gx\frac{\partial^2 z}{\partial x^2} + 2g\frac{\partial z}{\partial x} + 2g\frac{\partial z}{\partial x} + 2w\frac{\partial^2 z}{\partial x^2} = 0 \tag{3}$$

$$2b + 2c\left(\frac{\partial z}{\partial y}\right)^2 + 2cz\frac{\partial^2 z}{\partial y^2} + 2fy\frac{\partial^2 z}{\partial y^2} + 2f\frac{\partial z}{\partial y} + 2f\frac{\partial z}{\partial y} + 2gx\frac{\partial^2 z}{\partial y^2} + 2w\frac{\partial^2 z}{\partial y^2} = 0 \tag{4}$$

Upon inserting $x = 0$, $y = 0$, $z = 0$, $\frac{\partial z}{\partial x} = 0$, and $\frac{\partial z}{\partial y} = 0$ into Equations (3) and (4), the resultant computation manifests as Equation (5).

$$\begin{cases} \frac{\partial^2 z}{\partial x^2} = -\frac{a}{w} \\ \frac{\partial^2 z}{\partial y^2} = -\frac{b}{w} \end{cases} \tag{5}$$

Proceeding in a similar fashion, by differentiating Equation (2) first with respect to $x$ and then $y$, we obtain Equation (6).

$$\frac{\partial^2 z}{\partial x \partial y} = -\frac{h}{w} \tag{6}$$

By coupling Equations (5) and (6) with the first two terms of the Taylor series expansion, we derive Equation (7). Ultimately, the culmination of these calculations and derivations yields the result depicted in Equation (8).

$$f(x,y) = z = f(0,0) + x\frac{\partial z}{\partial x} + y\frac{\partial z}{\partial y} + \frac{1}{2}\left(x^2\frac{\partial^2 z}{\partial x^2} + 2xy\frac{\partial^2 z}{\partial x \partial y} + y^2\frac{\partial^2 z}{\partial y^2}\right) \tag{7}$$

$$z = \frac{1}{2}\left(-\frac{a}{w}x^2 - \frac{2hxy}{w} - \frac{b}{w}y^2\right) \tag{8}$$

Eccentricity is a function depicting the angular and spatial relationship between the axes of two cylindrical bodies. A higher eccentricity results in a broader contact surface between the two cylinders. The contact surface on the XY plane can be characterized using a quadratic surface, as described by Equation (8). Within this equation, parameters '$a$' and '$b$' govern the curvature of the contact surface in the X and Y directions, respectively, reflecting the width of the contact area, while '$h$' determines the rotation angle of the contact surface.

When the axes of the two cylinders are not perfectly parallel, an eccentricity distinct from zero emerges, implying that neither '*a*' nor '*b*' are zero. Consequently, the shape of the contact surface manifests as an ellipsoidal paraboloid. This suggests that the pressure distribution on the contact surface might be non-uniform, exhibiting peak values at the center and diminishing towards the edges.

Upon an in-depth examination of the contact surface geometry in conjunction with the actual operational conditions, it becomes evident that the clearance during the meshing process between the pinion teeth and the cycloid gear leads to the misalignment of the axes of the two contacting bodies. This misalignment culminates in an ellipsoidal paraboloid-shaped contact region. Moreover, the oscillation or 'runout' of the pinion teeth undeniably has repercussions on the stress distribution across the face of the cycloid gear, as shown in Figure 2.

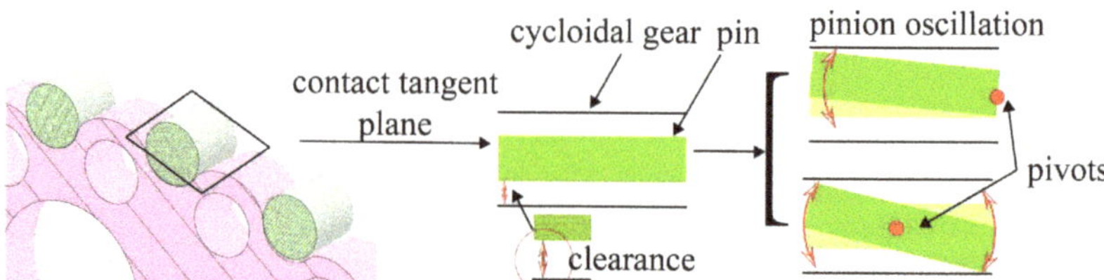


**Figure 2.** Non-parallel axis meshing of cycloid pin teeth.

In scenarios considering pinion oscillation, this analysis operates on the presumption of contact between infinitesimally small convex bodies. The pressure distribution on the contact surface, under these circumstances, still conforms to the Hertzian stress distribution model. Notably, we disregarded any contact effects beyond the direct meshing of the pinion with the cycloid gear. This means that during the 'runout' phase of the pinion, there are no other collisions to account for.

To sum up, when the pinion engages with the cycloid gear, the situation can be approximated as the contact problem of two cylindrical bodies with misaligned axes. The deformation at the contact site manifests as an ellipsoidal paraboloid. Hence, when computing the contact forces between the cycloid and the pinion, the impact of a conical contact surface on the force dynamics is incorporated into the analysis.

Assuming that the angle between the axes of two cylinders is $\theta$, where $\theta \in \left(0^\circ, 90^\circ\right]$, it should be noted that when $\theta$ is $0^\circ$, the relationship between the contact force and deformation between the cylinders is linear, and a different contact force model is used; hence, this value is not included in the range. In addition, the radii of the two cylinders are $R_1$ and $R_2$, respectively. It is possible to calculate the heights of the two surfaces near the contact point, as shown in Equation (9).

$$h = z_1 + z_2 = \frac{x^2}{2R_1} + \frac{(x\cos\theta - y\sin\theta)^2}{2R_2} \tag{9}$$

where $R_1$ and $R_2$ denote the radii of the two cylinders, respectively.

By forming a quadratic form through the surface height $h$, the principal curvatures can be obtained, and consequently, the results for the equivalent radius are derived. This allows for the substitution of the equivalent rigid sphere and elastic half-space body contact force model (which is identical to the computational model used when the angle between the axes is $90^\circ$), as shown in Equation (10).

$$F = \frac{4}{3}E^*\widetilde{R}^{1/2}d^{3/2}, \tag{10}$$

The aforementioned formulas are based on the assumption of small deformations and that the cylindrical bodies are made of linear elastic materials. Under the conditions of complete contact, these equations are derived from the Hertzian contact theory model. The contact force $F$ can be expressed as Equation (11).

$$F = \frac{4}{3}E^*(R_1R_2)^{1/4}d^{3/2}\cos^{3/2}\left(\frac{\pi - 2\theta}{4}\right) \tag{11}$$

where $E^*$ represents the equivalent elastic modulus, assuming that the material of the cycloid wheel and the pin gear is the same, both being GCr15, $E^* = \frac{0.5E}{1-v^2}$. $d$ represents the contact depth. $\cos^{3/2}\left(\frac{\pi - 2\theta}{4}\right)$ denotes the ratio of the long and short axes of the ellipse in the contact region, and it takes into account the effect of the angle between the two cylindrical axes on the shape of the contact surface.

## 3. Analysis Method of End Face Stress for Cycloid Gear

Various methodologies are employed for the computation of stress on gear contact surfaces, including Mechanical Analysis, FEA, Empirical Formulation, the equivalent stress method, and Statistical Analysis. The applicability of these approaches hinges on the specific scenarios they address. While Mechanical Analysis and FEA offer nuanced insights into the stress distribution and magnitude on the contact surface, they mandate a comprehensive understanding of the gear's structure and material properties, accompanied by intricate computational processes. The Empirical Formulation, albeit restricted in its applicability, proffers the advantages of simplicity and expeditious calculations. The equivalent stress method, despite its rapid computational attributes, has a circumscribed domain of applicability, potentially leading to discrepancies in its outputs. The Statistical Analysis approach is adept at mirroring the gear's lifecycle characteristics, but it is inherently sensitive to the volume and distribution of data samples.

### 3.1. Conformal Mapping Method for Multiply Connected Domains

The Schwarz–Christoffel (SC) transformation serves as a keystone in the realm of geometry, facilitating the conformal mapping of the upper half of the complex plane into the interior of polygons. Harnessing the potential of the SC transformation paves the way for the geometrical elucidation of complex functions, enabling a probe into their inherent properties and addressing the boundary value challenges that are prevalent in the spheres of physics and engineering [25,26]. A quintessential representation of the SC formula is typically articulated as follows:

$$\omega = f(z) = C \int^{z_0} \prod_{k=1}^{n} (z - z_k)^{a_k - 1}dz + C_1, \tag{12}$$

where the original vertices of the polygon $\omega_k$ correspond to the vertices $z_k$ after being mapped onto the circle. At the point $\omega_k$, the turning angle of the tangent to the polygon is $(1 - a_k)\pi$. The constants $C$ and $C_1$ are set based on the physical boundary conditions.

While the Schwarz–Christoffel (SC) mapping adeptly maps bounded polygons to the unit circle, its utility is confined to simply connected polygonal domains. When addressing bounded multiply connected polygonal regions, Mohamed et al. [27,28] have elucidated an approach grounded in the Koebe iterative method combined with boundary integral equations to ascertain mapping functions and their inverses for these multiply connected domains. This methodology furnishes a numerical implementation to compute circular mappings of multiply connected regions, leveraging the Koebe iterative procedure to approximate the mapping relations.

By discretizing the boundary components and deploying computationally economical algorithms, accurate circular mapping outcomes can be efficiently derived. Specifically, by employing the boundary values gleaned from the Koebe iterative method for bounded multiply connected regions, the mapping values for the interior points of the multiply connected domains can be determined using the Cauchy integral formula, as delineated in

Equation (13). In a parallel vein, the Cauchy integral formula facilitates the computation of the values for circular inverse mapping, as presented in Equation (14).

$$w = \omega(z) = \frac{1}{2\pi i} \int_J \frac{\omega(\eta(t))}{\eta(t) - z} \eta'(t) \mathrm{d}t, \tag{13}$$

where $w$ is the value of the mapping function, representing the point on the complex plane after mapping; $z$ is the point on the complex plane waiting to be mapped; $\omega(\eta(t))$ denotes the value of the mapping function $\omega$ on the parameter curve $\eta(t)$; $\eta(t)$ is the parameter curve describing the path in the bounded region $J$; and $\eta'(t)$ represents the derivative of the parameter curve $\eta(t)$ with respect to $t$.

$$z = \omega^{-1}(w) = \frac{1}{2\pi i} \int_J \frac{\omega^{-1}(\xi(t))}{\xi(t) - w} \xi'(t) \mathrm{d}t, \tag{14}$$

where $z$ is the value of the inverse mapping function, indicating the point mapped to the complex plane; $w$ is the point on the complex plane waiting to undergo inverse mapping; $\omega^{-1}(\xi(t)))$ signifies the value of the inverse mapping function $\omega^{-1}$ on the parameter curve $\xi(t)$; $\xi(t)$ is the parameter curve that outlines the trajectory in the bounded region $J$; and $\xi'(t)$ is the derivative of the parameter curve $\xi(t)$ with respect to a given parameter.

### 3.2. Geometric Characteristics of Cycloid Gear Profiles with Pores

The maximum allowable diameter of the cycloidal gear pinhole is represented by Equation (15). The corresponding pin hole structure arrangement is shown in Figure 3.

$$[D_{AM}] = \min \left\{ \begin{array}{c} D_{PH} - D_{CI} - 2 \times 0.03D_C \\ D_{PH} \sin \frac{\pi}{z_w} - 0.03D_C \end{array} \right\}, \tag{15}$$

where $D_C$ signifies the diameter of the cycloidal gear, $D_{PH}$ represents the diameter of the center circle of the cycloidal gear pinhole; and $D_{CI}$ denotes the diameter of the central hole of the cycloidal gear, and in this case, the value of $D_{CI}$ is set to $0.4D_C$. It can also be determined based on the actual outer diameter of the installed bearing. The variable $0.03D_C$ indicates the empirical minimum wall thickness values for the pin holes.

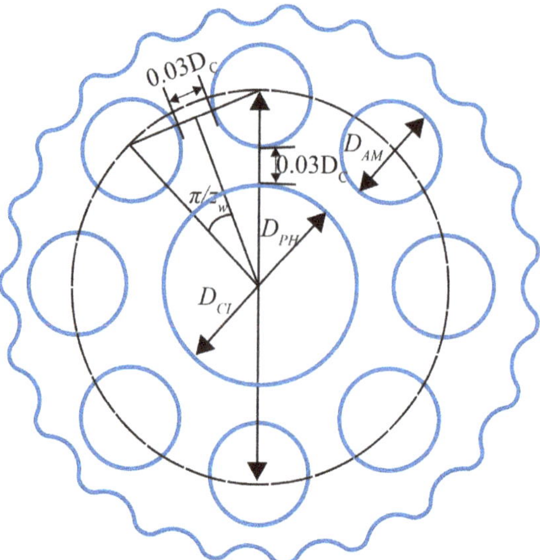


**Figure 3.** Arrangement of cylindrical pin holes.

### 3.3. SACGES-CMMD Method

By utilizing the dual quaternion cycloid gear tooth profile model delineated by Jiang [29], a collection of vertex coordinates was garnered. With the aid of the Plgcirmap toolbox, the bounded connected polygonal domain was adeptly transformed into a bounded connected unit circle domain. Subsequent to this transformation, boundary vertices were systematically numbered.

During the meshing phase, the areas of the cycloid gear teeth subject to force warranted a localized refinement. Consequently, a mesh gradient of 1.5 was set, and the GeometricOrder was designated as 'quadratic'. For the boundary conditions, the nodes associated with the central hole were anchored with fixed constraints. Following these configurations, the stress analysis results of the unit circle were then deduced. Ultimately, the coordinates endowed with stress details were inversely mapped from the unit circle back to the polygonal domain. The procedural workflow of the stress analysis method of the cycloid gear end surface based on conformal mapping in multi-connected domains (SACGES-CMMD) is vividly depicted in Figure 4.

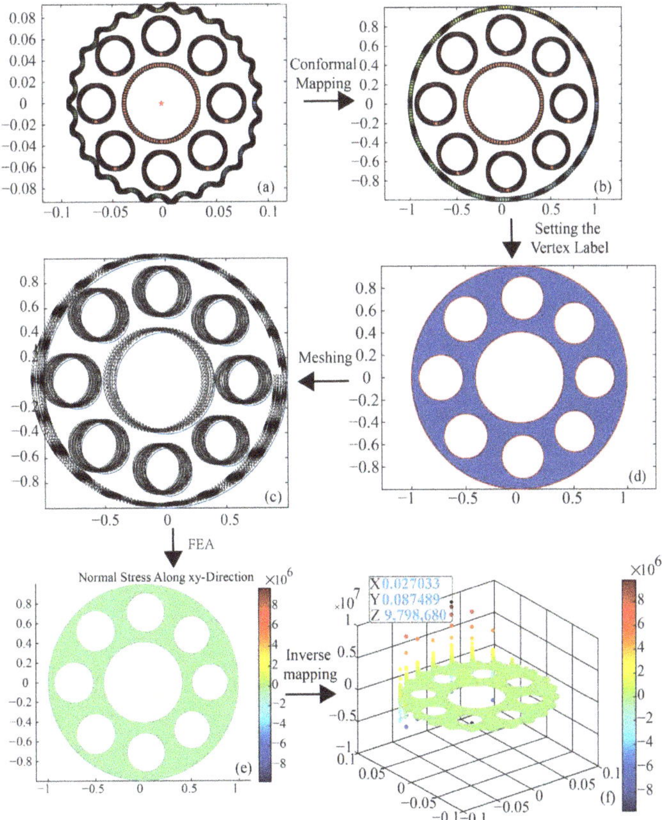


**Figure 4.** SACGES-CMMD workflow. (**a**) Cycloidal gear profile polygon; (**b**) Unit circle; (**c**) A unit circle with vertex labeling information; (**d**) Unit circle mesh division; (**e**) Unit circle shear stress results; (**f**) Cycloidal gear end face stress results. By amalgamating the Boundary Integral Method and Koebe's iterative technique, this research ventured into computations employing hierarchical grid points, the Nystrom approach, and the GMRES solver. The method undertakes the discretization and resolution of polygons via boundary integral equations, capitalizing on the generalized Neumann kernel, and thereby facilitating the derivation of numerical results for both the mapping function and its inverse.

In the framework of this analysis, a bounded polygonal connected domain boasting a connectivity value of 10 was explored. To capture the conformal mapping of this domain, the 'plgcirmap' function in MATLAB 2020b was employed, leveraging the vertices of the polygon and selected points within the domain. During the mapping computation, normalization was invoked. The resultant conformal mapping, $f$, and its inverse, $f^{-1}$, were pictorially represented using the 'plotmap' function, as evinced in Figure 5.

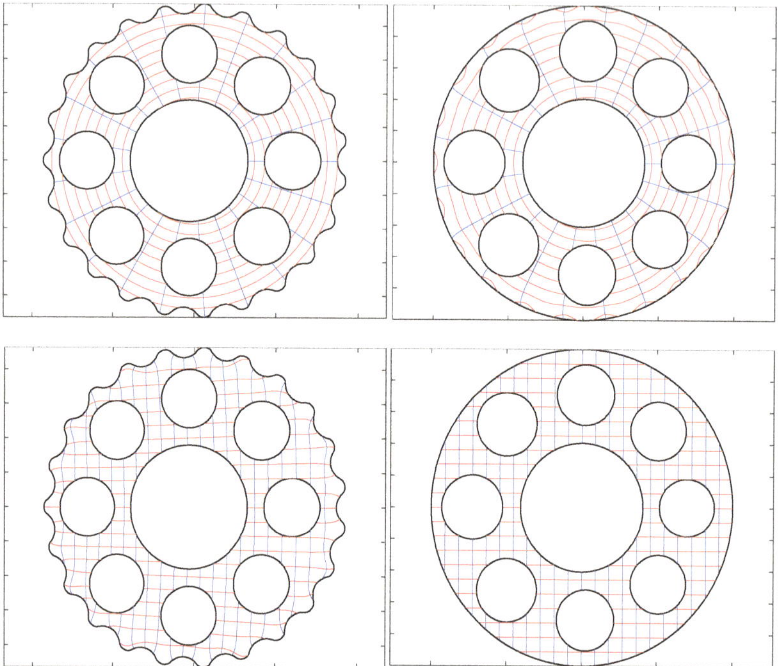

**Figure 5.** Bounded multiconnected domain mapping results for cycloid contours.

In this investigation, an intricate mapping procedure was carried out, converting the node information from a bounded multiply connected region, specifically the epitrochoid area, to a unit circle. Building on this, the MATLAB PDE toolbox was employed to execute a stress analysis on the unit circle, delineating the process from stages (c) to (d), as shown in Figure 4. The subsequent inverse mapping process from the unit circle back to the polygonal region, embedding the stress information, necessitates the computation of the Jacobian matrix, executed as follows:

(1)  Initialization: For a designated polygonal region coupled with its associated stress tensor distribution, the position of each node (polygon vertex) and its corresponding first-order derivative are computed. Special attention is devoted to instances where the first-order derivative is zero, circumventing potential numerical complications in ensuing computations.

(2)  Computation of Jacobian Matrix for Inverse Mapping: Upon obtaining the node data and first-order derivatives, the Jacobian matrix for inverse mapping at each node is ascertained. The Jacobian is informed by the complex derivative of the inverse mapping function. By inputting complex variable details representing the unit circle node into the inverse mapping function, the correlated polygon node data are derived, and the differential from this reverse mapping (from the unit circle to the polygon) is secured. When constructing the Jacobian matrix, the derivative of the complex variable, normalized by its magnitude, is bifurcated into real and imaginary parts,

serving as matrix elements. This protocol adheres to the Cauchy–Riemann equations characterizing complex functions. The result functions as a local Jacobian matrix for rotation and scaling at the node.

(3)   Mapping of Stress Tensor: During inverse mapping, the stress tensor on the unit circle is projected to nodes within the polygonal region. The need for scaling stems from the disparities in the dimensional magnitudes across the polygon. To accurately capture the stress levels, especially given the diverse stress distribution in various regions, a uniform global scaling factor is introduced based on the specific contour dimensions of the shape. This ensures a more authentic representation of the stress distribution throughout the domain. Additionally, each node might undergo local rotation and scaling, as described in step 2. These transformations are assessed in relation to the local node coordinate system, necessitating individual node calculations. When integrating these local adjustments with the global stress scaling influenced by dimensional magnitudes, a holistic and precise depiction of the stress distribution within the polygon emerges. It is crucial to balance the considerations of this uniform global scaling factor with the local rotational and scaling intricacies during stress tensor mapping within the polygonal domain. This study not only underscores this integration but also dives deep into the implications of the global scaling factor, laying the groundwork for subsequent inquiries in this domain.

(4)   Storing Mapping Outcomes: The stress tensor derived from mapping is archived within a novel matrix representing the stress distribution within the unit circle, culminating the tensor mapping from the polygonal to the unit circle domain.

### 3.4. Global Stress Scaling Factor

Considering the size discrepancy at the millimeter scale between the shape and the unit circle, a global stress scaling factor is employed in this paper for normalization purposes. As depicted in Figure 6, by holding the distance between the fixed point and the point of applied force constant, the alterations in the shape of the force-applied region have negligible impacts on the resultant maximum equivalent stress. When ensuring consistency in the mesh count, as per the representation in Figure 7, an inverse relationship between the maximum equivalent stress value and side length is evident. Their product, however, demonstrates minimal fluctuations. From the above observations, during the mapping process, it is discerned that the factors affecting the peak stress predominantly hinge on variations in the geometric dimensions, with shape deviations playing marginal roles.

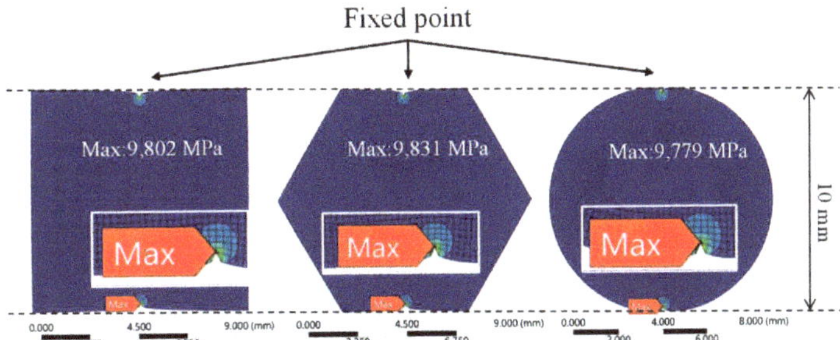


**Figure 6.** Influence of geometry on maximum equivalent stress (subjected to a force of 100 N vertically upwards with a mesh size of 0.1 mm).

In light of these findings, this paper introduces a methodology for computing the average radius of a polygon centered on its geometric centroid, which essentially serves to determine the scaling factor. The initial step merges the horizontal and vertical coordinates of each vertex of the polygon into a singular matrix. Subsequently, the geometric centroid of the polygon is ascertained. The ensuing step involves calculating the Euclidean distances from each vertex to the geometric centroid. The average of these distances renders the polygon's average radius. The core intent behind this strategy is to harness the average radius as a pivotal reference value during geometric scaling transformations.

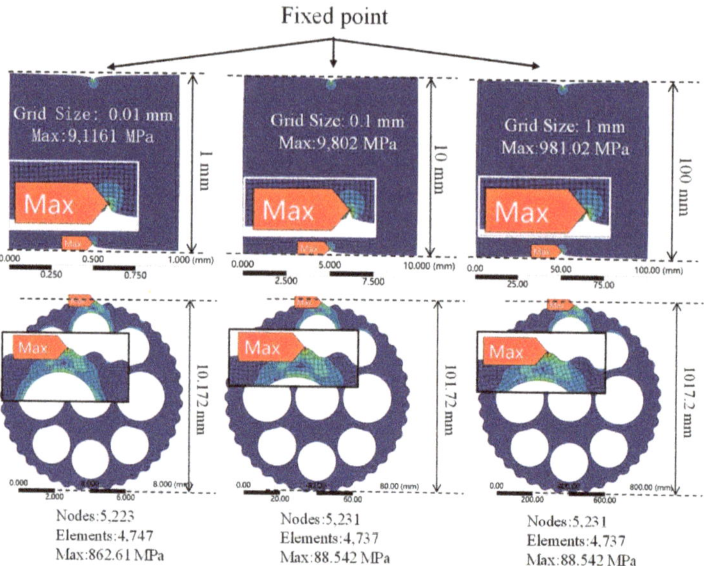



**Figure 7.** Influence of maximum size on maximum equivalent stress (subjected to a force of 100 N vertically upwards).

For a more formal representation, consider $P$ as a two-dimensional polygon comprising $n$ points, with the coordinates for each point delineated as $(x\_i, y\_i)$, where $i = 1, 2, \ldots, n$. Consequently, the average radius $r\_avg$ of the polygon $P$ can be articulated as per Equation (16).

$$r_{\mathrm{avg}} = \frac{1}{n}\sum_{i=1}^{n}\sqrt{\left(x_i - \frac{1}{n}\sum_{j=1}^{n} x_j\right)^2 + \left(y_i - \frac{1}{n}\sum_{k=1}^{n} y_k\right)^2}. \tag{16}$$

Given the context established in this study, the stress transformation process is delineated from unit circle inversion mapping to the polygon. Hence, the scaling factor is determined as $1/r\_avg$. Assuming that, during the variations in the geometric shape dimensions, the product of the stress and the scaling factor remains constant, this approach facilitates the determination of the stress across the cycloid gear cross section.

## 4. Bayesian-Enhanced Least Squares Genetic Algorithm Optimization Method

When integrating the research contents of Sections 2 and 3, there is a need for an optimization algorithm to address the stress distribution optimization problem, considering the errors and random fluctuations in the contact force. In this study, we introduce the BLSGAOM, an innovative approach that seamlessly integrates Genetic Algorithms (GAs) with Least Squares Boosting Models, as shown in Algorithm 1. Initially, GA parameters are established within specified bounds, after which an iterative optimization, steered by BLSGAOM's fitness function, adapts and fine-tunes these parameters. The specific settings for the input parameters are as follows: in the authors' previous study [29], three error values were identified as the input parameters, with their respective numeric bounds defined as [0 mm, 0.1 mm], [0 mm, 0.04 mm], and an unspecified range denoted by [0′, 10′]. The three represent the bending deformation of the input crankshaft, the amount of pinion runout, and the hysteresis angle, respectively.

---

**Algorithm 1** Bayesian-enhanced least squares genetic algorithm optimization model

**Step Procedure**

1  **Set Genetic Algorithm Parameters:**
   nvars← 3
   **lb**← $[0, 0, 0]$
   **ub**← $[0.1, 0.04, 10/60]$

2  **Initialize Genetic Algorithm Options:**
   options ← SET Display to 'iter', Migration Interval to 20, Migration Fraction to 0.3
   **bestFvals**←Φ

3  (x, fval) ← GA using **fitnessFunction**, lb, ub, and options

4  Display optimal solution and fitness: $x, fval$

5  **Plot evolution of Best Fitness Value:**
   PLOT bestFvals **vs.** Generation

6  **Function** fitnessFunction(x):
   $y_{pred}$ ← LSBM_**predict**_result($x_1, x_2, x_3$)
   **return** $y_{pred}$

7  **Function** outfun(options, state, flag): UPDATE bestFvals with min(state.Score)
   DISPLAY state's details
   **return** options, state, optchanged

8  **Function** LSBM_predict_result($x_1, x_2, x_3$): LOAD pretrained model
   $X$ ← $[x_1, \mathbf{x_2}, x_3]$
   result ← **PREDICT** with model and $X$
   **return result**

9  LOAD Dataset
   Extract input X and **target** y Split Dataset using cvpartition

10  **Hyperparameter Tuning:** Set optimization options results ← BAYESOPT

11  Train bestModel using FIT with optimal parameters

12  Predict: $y_{pred}$ ← PREDICT using bestModel and X

13  Compute metrics: MSE, RMSE, MAE, $R^2$

14  Save trained model

15  Plot Actual vs. Predicted

---

A salient feature of our BLSGAOM is its adept utilization of surrogate models during the Bayesian optimization phase. These surrogate models have become indispensable tools in engineering design, offering substantial advantages, especially when faced with computationally challenging tasks. The incorporation of surrogate models offers a dual benefit: they significantly reduce the computational overhead by efficiently approximating complex problem behaviors and provide flexibility, enabling the seamless application of classical optimization algorithms. This malleability helps to sidestep potential pitfalls such as local optima and high computational costs, which are often inherent to the original problem.

Moreover, surrogate models enhance the interpretability of the foundational problem, making them invaluable in contexts where understanding the nuances of the problem is crucial. In scenarios characterized by expansive or intricate design spaces, these models proficiently navigate, identifying solutions that best align with specific criteria. Their inherent adaptability means they can easily assimilate new data, ensuring continuous refinement in their precision and efficiency.

By leveraging the strengths of these surrogate models, BLSGAOM provides a judicious approximation to the true objective function. This methodological approach ensures thorough exploration and exploitation of the hyperparameter space without resorting to excessive evaluations of the primary objective. The culmination of this process is a refined model armed with optimal settings. To validate the robustness of our model, the predictions are critically assessed using a suite of error metrics, providing a holistic view of its performance and reliability. The optimized model is shown in Equation (17).

$$\text{minimize} \quad f(b,f,d) = \max(\sigma(b,f,d)) + \text{var}(\sigma_{\text{dist}}(b,f,d))$$
$$\text{s.t.} \quad 0 \le \Delta_b \le 0.1$$
$$0 \le \Delta_f \le 0.04 \tag{17}$$
$$0' \le \Delta_d \le 10'$$

where $f(b,f,d)$ represents the objective function of optimization; $\sigma(b,f,d)$ represents the maximum stress induced by the error terms; and $\sigma_{\text{dist}}(b,f,d)$ represents the stress distribution caused by the error terms.

## 5. Results and Discussion

### 5.1. Results of Clearance Effect on Contact Force Fluctuation

For relatively minute angular deviations, as illustrated in Figure 2, the range of angular values can be approximated to the ratio of the clearance to pinion length and half of the clearance over the pinion length. Specifically, when the clearances are set at 10 μm, 20 μm, and 30 μm, respectively, they correspond to the angular variations of [0.001 rad, 0.002 rad], [0.002 rad, 0.004 rad], and [0.003 rad, 0.006 rad]. Within these bounds, the relationship between the contact force and contact depth is depicted in Figure 8. Post the establishment of a slight angular discrepancy due to the clearance, there is an overall reduction in the contact force. Additionally, as the clearance marginally increases within a certain range, there is a subtle rise in the contact force.

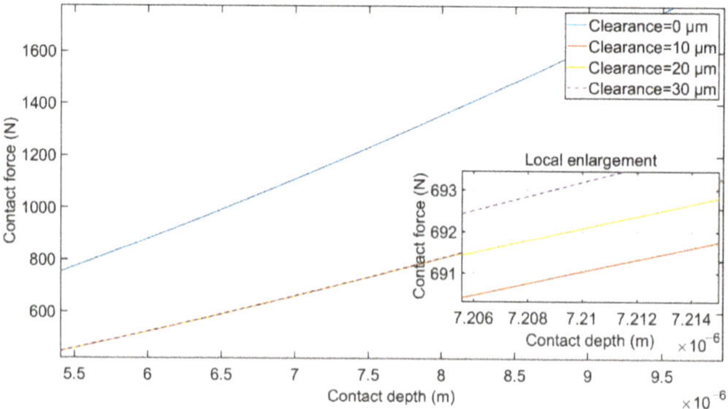


**Figure 8.** Relationship between contact force and contact depth for different clearance values.

### 5.2. Accuracy Results for Conformal Mapping

In the present study, a meticulous computation and an analysis were conducted on the errors at a series of test points for complex functions. Initially, a set of test points were uniformly sampled on the unit disk, and through the inverse mapping of the complex functions, these points were correspondingly mapped to a collection on the complex plane, as illustrated in Figure 9a. Subsequently, the discrepancies between each test point and its associated point post inverse mapping were evaluated, with the maximum norm employed as the metric for error quantification.

According to the computed outcomes, a parabolic trend in the distribution of errors between the test points and their inversely mapped counterparts was discerned, as depicted in Figure 9b. Within this error distribution, the maximum error norm was found to be $2.7280 \times 10^{-11}$, signifying that the inverse mapping method exhibited exemplary performance in the majority of instances. Such a method could accurately preserve the relationship between the test points and the original points, underscoring its robustness and precision.

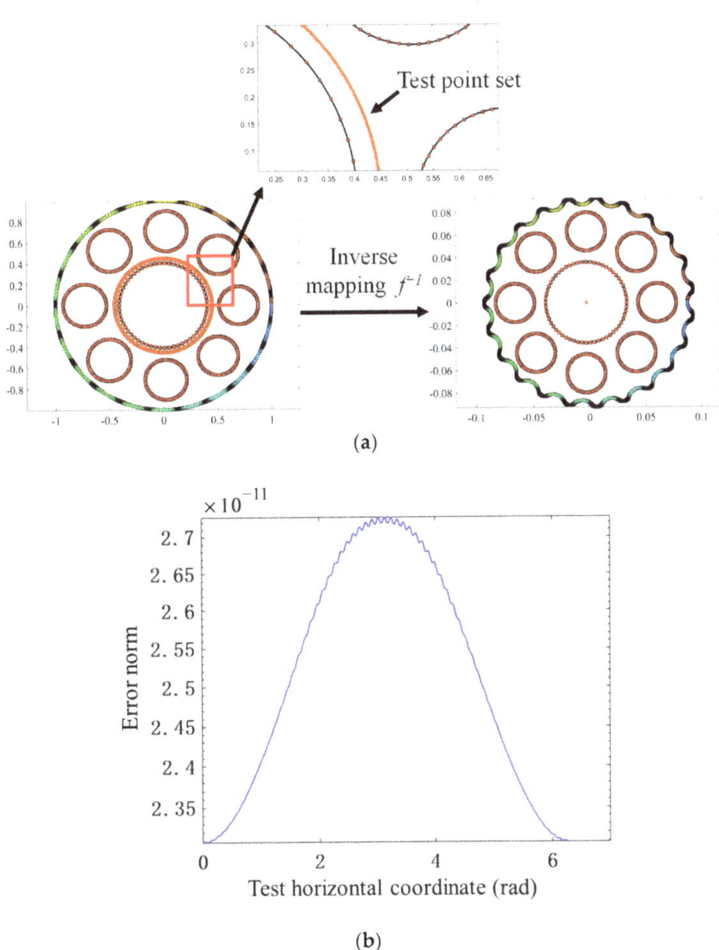



**Figure 9.** Accuracy results for conformal mapping. (**a**) Test point set. (**b**) Variation of the error norm during the inverse mapping process.

### 5.3. SACGES-CMMD vs. FEM Comparison Results

The SACGES-CMMD, in contrast to the FEM, brings forward a set of unique advantages. Central to these is the conformal mapping technique, which preserves angular relationships within multi-connected domains. This ensures that the stress patterns on a mapped unit circle accurately reflect those of the original polygonal domain. This geometric preservation, coupled with a scaling factor, streamlines stress computations. Additionally, this method boasts greater numerical stability, minimizing the potential instabilities and singularities that are often associated with direct FEM calculations on polygons. Moreover, applying the FEM on the unit circle guarantees a uniformly distributed grid, benefiting from the circular domain's inherent regularity and thereby heightening the computational precision. Furthermore, the straightforward circular geometry of the stress distribution facilitates an enhanced visualization and comparability of the results. Collectively, the use of conformal mapping for stress distribution in multi-connected domains enhances geometric integrity, numerical robustness, grid uniformity, and the clarity of the visualization and analysis.

From a comparative analysis of the results obtained using the FEM and SACGES-CMMD, as depicted in Figure 10 and summarized in Table 1, four key observations can be drawn:

(1) Grid Scale and Gradient Consistency: Under equivalent computational conditions, both methodologies exhibited grid sizes and gradients of comparable scales.

(2) Optimized Grid Utilization with SACGES-CMMD: Notably, the SACGES-CMMD approach utilized marginally fewer grid nodes than its FEM counterpart. The smallest reduction ratio in the number of nodes was 7.34%. As the mesh size increased, the reduction ratio in the nodes gradually increased as well, reaching up to 19.34% in the end. This suggests that the conformal mapping process underpinning SACGES-CMMD permits the elimination of superfluous grid elements, thereby optimizing computational efficiency within a compact grid layout.

(3) Comparative Stress Metrics: The comparison highlighted a close correspondence between the maximal shear stress and the stress distribution's STD for both methods, with no significant disparities discernible.

(4) Computational Efficiency: While the computation durations of both techniques were roughly commensurate, a downward trend in the grid element count conferred the SACGES-CMMD with a slight edge in the computational expediency. In the comparison of the computational efficiency, when the mesh was dense, both exhibited similar efficiencies with variations of 0.11% and 0.29%, respectively. As the mesh size increased, the computational efficiency of SACGES-CMMD showed a significant improvement, reaching a peak of 4.75%.

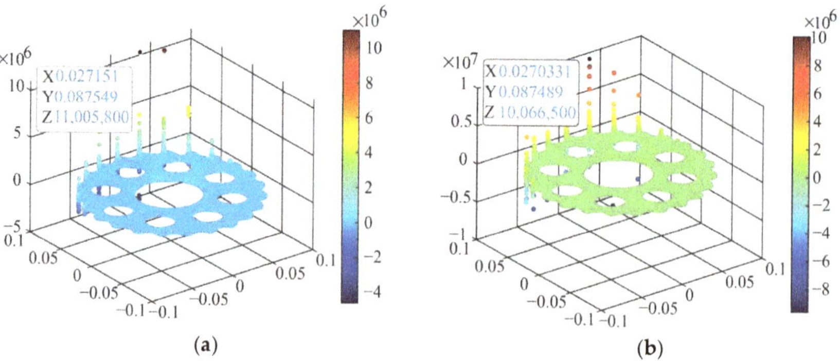



**Figure 10.** Comparison of shear stress analysis results. (**a**) FEM shear stress distribution (scaling factor of 0.0892); (**b**) SACGES-CMMD shear stress distribution (scaling factor of 0.0892).

**Table 1.** Comparison of FEM and SACGES-CMMD results.

| | Scaling Factor | Maximum Mesh Size (mm) | Minimum Mesh Size (mm) | Mesh Gradient | Number of Nodes | Percentage Reduction (%) | Maximum Shear Stress (MPa) | Stress Distribution STD | Computation Time (s) | Percentage Reduction (%) |
|---|---|---|---|---|---|---|---|---|---|---|
| FEM | 0.0892 | 0.892 | 0.0892 | | 98,202 | | 11.01 | 0.1028 | 18.95 | |
| SACGES-CMMD | | 10 | 1 | | 90,992 | 7.34 | 10.07 | 0.1275 | 18.97 | −0.11 |
| FEM | 0.0898 | 1.3470 | 0.0988 | | 55,896 | | 10.65 | 0.1296 | 17.21 | |
| SACGES-CMMD | | 15 | 1.1 | 1.5 | 49,020 | 12.31 | 10.09 | 0.1749 | 17.26 | −0.29 |
| FEM | 0.0900 | 1.8 | 0.0990 | | 46,758 | | 10.72 | 0.1421 | 17.03 | |
| SACGES-CMMD | | 20 | 1.1 | | 39,451 | 15.63 | 10.12 | 0.1931 | 16.22 | 4.75 |
| FEM | 0.0904 | 2.712 | 0.0994 | | 38,654 | | 10.68 | 0.1555 | 16.56 | |
| SACGES-CMMD | | 30 | 1.1 | | 31,178 | 19.34 | 9.47 | 0.2135 | 15.99 | 3.45 |

When synthesizing these insights, it becomes evident that while SACGES-CMMD parallels the FEM in terms of computational precision and reliability, it proffers certain advantages. Namely, these comprise a leaner grid element composition and a modest reduction in the computation time. Consequently, SACGES-CMMD emerges as a viable strategy for stress distribution evaluations in multi-connected domains, holding potential for elucidating the characteristic stress distribution patterns within such domains. This stress analysis method provides a thought process for model equivalence transformation to solve the issue of computational infeasibility of the solid model before and after the cycloidal wheel modification across large-scale grid divisions. This is due to the fact that the amount of modification is at the micron level, while the tooth profile shape is at the millimeter level. It offers a solution for large-scale grid divisions in computational models.

### 5.4. Optimization Results of Stress Distribution Based on BLSGAOM

Drawing upon the parameter design presented in Table 2, the contact force dynamics of the gear tooth profile are dissected. This analytical framework then sets the stage for the integration of conformal mapping techniques specific to multi-connected domains. The intricate cycloidal gear tooth profile is seamlessly mapped onto a unit circle through this approach.

**Table 2.** Design parameters of the cycloid-pin system for SACGES-CMMD.

| Parameters | Values | Parameters | Values |
|---|---|---|---|
| Loading torque, $T$ (Nm) | 600 | Number of teeth of cycloid gears, $z_a$ | 23 |
| Density, $\rho$ (Kg/m$^3$) | 7830 | Pinion number, $z_b$ | 24 |
| Poisson's ratio, $\nu$ | 0.3 | Pitch radius of pin teeth, $r_b$ (mm) | 72 |
| Pin tooth center circle radius, $R_b$ (mm) | 100 | Pitch radius of the cycloid gear, $r_a$ (mm) | 69 |
| Eccentricity, $e$ (mm) | 3 | Pin tooth radius, $r_p$ (mm) | 10 |
| Elastic modulus, $E$ (GPa) | 208 | | |

The magic of this conformal mapping lies in its transformative capability; it simplifies the complex contour of the gear tooth to the elementary geometry of a unit circle. As alluded to in Section 3.3, utilizing the approximate relationships of the scaling factors enables the stress distribution results, which were originally derived from the FEM on the unit circle, to be inversely mapped onto the cycloidal gear tooth profile. This technique substantially curtails the computational overhead that is typically associated with stress distribution calculations, offering both efficiency and precision.

For this study, the Least Squares Boosting (LSBoost) method was enlisted to train the machine learning models. This endeavor commenced with the segregation of the dataset; of the 1000 data instances, 80% fueled the training phase, while the remaining 20% were earmarked for testing. Two pivotal hyperparameters—the learning rate and the number of learning epochs—were subjected to rigorous optimization. Specifically, the learning rate oscillated between 0.01 and 0.2, whereas the epochs ranged from 50 to 200. The Bayesian optimization, as depicted in Figure 11, underpinned the hyperparameter tuning phase, selecting the optimum values based on the anticipated improvements. The k-fold cross-validation technique was employed during model training to ensure both a robust assessment of the performance and a bulwark against overfitting. Upon finalizing the hyperparameters, the model was trained using the dataset and subsequently engaged in predictions.

To rigorously evaluate the predictive capacity of the model, we employed a comprehensive set of critical performance metrics. The Mean Square Error (MSE), a key indicator of the predictive accuracy, denoted an average squared difference of $9.6527 \times 10^{-7}$ between the predicted and observed values, highlighting the model's precision. This precision was further reinforced by the Root Mean Square Error (RMSE), valued at $9.8248 \times 10^{-4}$, and the Mean Absolute Error (MAE), which reflected an average deviation of $2.7699 \times 10^{-4}$. Additionally, the model's explicative prowess was exemplified by an R-squared value of 0.9998, indicating that it could account for nearly all the variability in the data, as graphically illustrated in Figure 12. In pursuit of sustained research continuity and replicability, the optimally trained model was archived, priming it for prognosticating fresh datasets. This strategic archiving not only ensures seamless future applicability but also fosters the study's replicability in analogous scenarios.

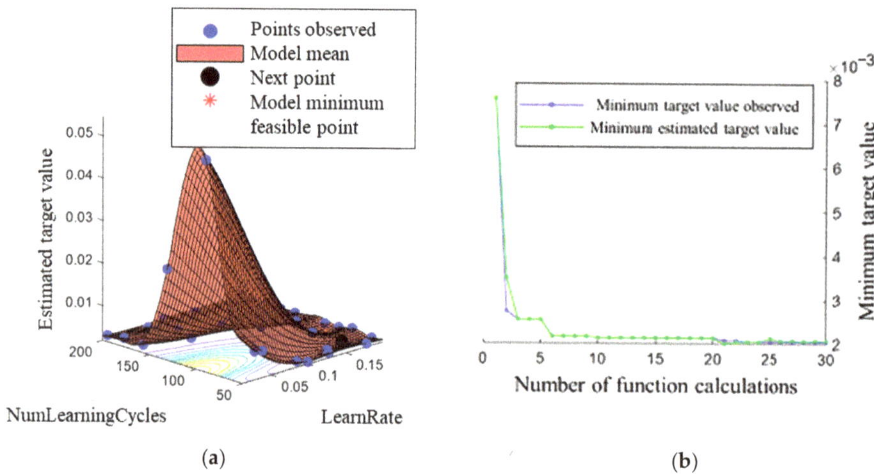



**Figure 11.** Variation of objective function model and minimum objective value during Bayesian optimization. (**a**) Objective function model. (**b**) Minimum objective value vs. number of function evaluations.

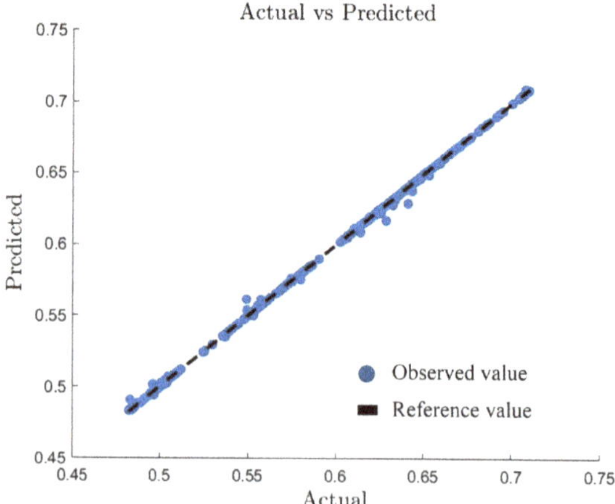



**Figure 12.** Predicted value vs. actual value.

In the present investigation, the GA was harnessed for model parameter optimization. Drawing inspiration from the principles of natural selection and genetic mechanics prevalent in biological evolution, the GA emerges as a heuristic global optimization approach renowned for its exemplary global search capabilities.

The predictive model derived earlier was amalgamated with the three error terms, serving as the basis to forecast outcomes based on the input parameters. Thus, the predictions that were generated constituted the fitness function for the GA, with the overarching objective being the identification of the input parameters that optimize (minimize in this context) the aggregate of the maximum stress and the stress distribution STD.

In this study, the configuration of the GA parameters was elucidated as follows: The solution vector length was set at three, directly corresponding to the three error terms. The defined search space bounds ranged from a lower boundary of [0 mm, 0 mm, 0°] to an upper boundary of [0.1 mm, 0.04 mm, 10/60°]. The output functions were integrated to record the optimal fitness values upon the conclusion of every generation. To bolster population diversity and sidestep the potential challenges of the local optima, migrations were programmed at intervals of every 20 generations, with each migration facilitating a randomized transfer of 30% of the population. When the GA was executed, the resulting optimal solution was pinpointed at [0.0641 mm, 0.0058 mm, 0.0584°]. This solution corresponded to a peak fitness (predictive error) value of 0.4691. A graphical representation depicting the progression of the optimal fitness value across generational transitions can be found in Figure 13.

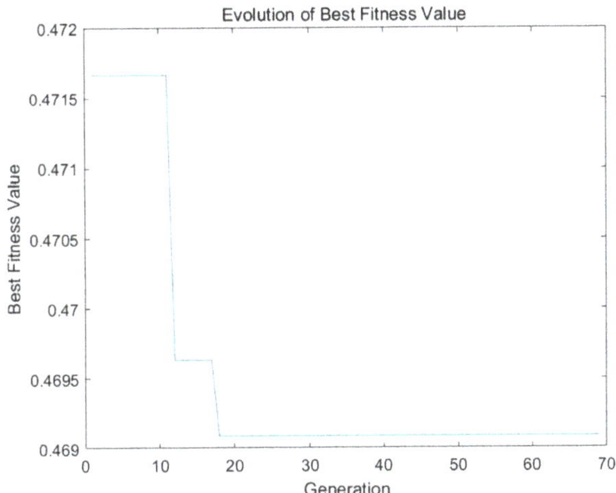

**Figure 13.** Evolution process of optimal fitness value.

### 5.5. Optimization Results of Stress Distribution Considering Contact Force Fluctuations

In the assembly, deviations manifesting as gear tooth jitters emerge due to the assembly clearances, subsequently leading to alterations in the contact forces. As stipulated by Equation (11), a rectification term, $\cos^{3/2}\left(\frac{\pi-2\theta}{4}\right)$, is incorporated during the computation of the contact forces. As with the parameter configurations delineated in Section 5.4, the optimization trajectory is vividly portrayed in Figure 14.

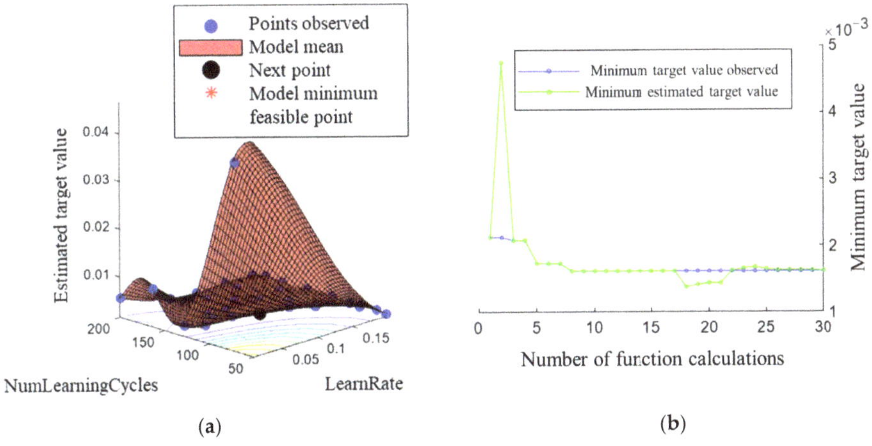


**Figure 14.** Variation of objective function model and minimum objective value considering contact force fluctuations. (**a**) Objective function model. (**b**) Minimum objective value vs. number of function evaluations.

Ultimately, the performance metrics of the predictive model were elucidated as $MSE = 6.8474 \times 10^{-8}$, $RMSE = 2.6168 \times 10^{-4}$, and $MAE = 4.6979 \times 10^{-5}$, and the coefficient of determination, $R^2$, stood at 1. Collectively, these indices underscore the model's robust explanatory power concerning the data, as graphically elucidated in Figure 15.

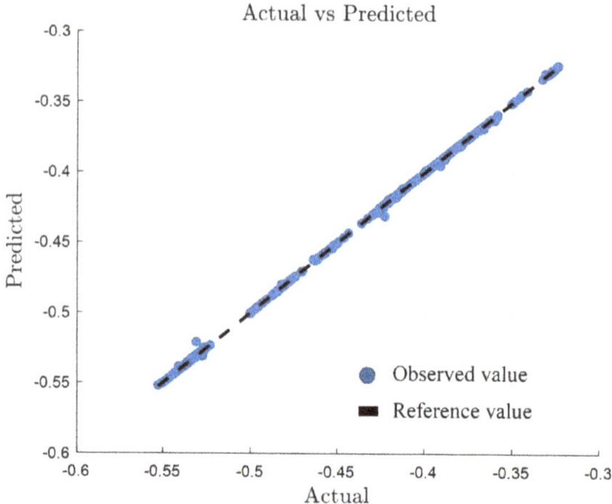

**Figure 15.** Predicted value vs. actual value under the influence of contact force fluctuations.

Upon the execution of the genetic algorithm, the optimal solution procured was [0.0645 mm, 0.0083 mm, 0.0819°]. This corresponds to an optimal fitness value (i.e., predictive error) of 0.4648. The trend of this optimal fitness value in relation to the number of generations can be visually represented in Figure 16.

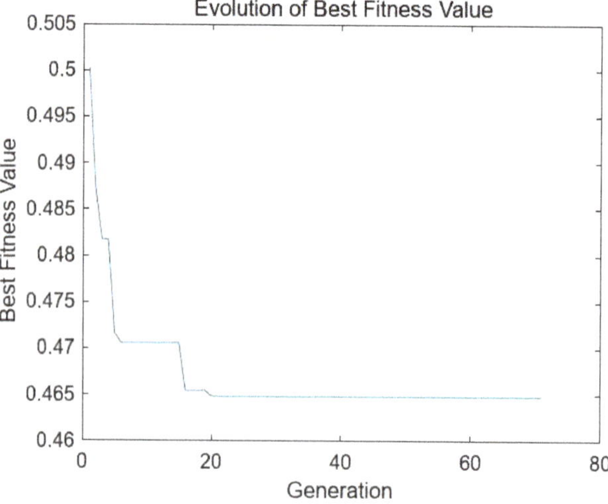

**Figure 16.** Evolution process of optimal fitness value under the influence of contact force fluctuations.

In the comprehensive analysis on the stress distribution in the cycloid-pin gear systems, two distinct methodologies were adopted: one that factored in the nuances of the contact force fluctuations and another that disregarded these fluctuations. Before incorporating the effect of the contact force fluctuations, the Bayesian-enhanced least squares boosted surrogate model exhibited an R-squared value of 0.9998, an MSE of $9.6527 \times 10^{-7}$, an

RMSE of $9.8248 \times 10^{-4}$, and an MAE of $2.7699 \times 10^{-4}$. This model predicted an optimal error configuration of [0.0641 mm, 0.0058 mm, 0.0584°] with a predictive error (or fitness value) of 0.4691.

Upon the consideration of contact force fluctuations, brought about by gear tooth jitters due to assembly clearances, the model's performance metrics exhibited an even higher fidelity. The R-squared value reached perfection, standing at 1, while the MSE, RMSE, and MAE values were reduced to $6.8474 \times 10^{-8}$, $2.6168 \times 10^{-4}$, and $4.6979 \times 10^{-5}$, respectively. The optimal error values in this scenario were [0.0645 mm, 0.0083 mm, 0.0819°], corresponding to a slightly improved predictive error of 0.4648.

The observed improvements in the model's performance metrics when accounting for the contact force fluctuations underpin the significance of considering such dynamic effects in the optimization process. By ensuring a holistic view of the system behavior, this study provides a pivotal foundation for more accurate predictive modeling and optimization in gear systems.

## 6. Conclusions

In this exploration to optimize the end face stress distribution of cycloidal gears, a parabolic contact surface between two cylindrical bodies was deduced, leading to the establishment of a model relating the contact force to the axis angle (between two axes) under varying clearances, and obtaining the corresponding impact factor. Concurrently, the adoption of the conformal mapping technique facilitated the geometric and stress transitions between the polygons and a unit circle. By harnessing the intricate multi-connected domain tooth profile of the cycloidal wheel, the approach synergized conformal mapping with a global stress scaling factor, promoting a reverse mapping from the unit circle to the polygons, and thereby determining the end face stress of the cycloidal wheel. The SACGES-CMMD methodology, compared to the FEM, showcased fewer grid divisions with a reduction ratio reaching up to 19.34%, improved computational efficiency peaking at 4.75%, and retained precision. To conclude, the employment of the Bayesian-enhanced least squares surrogate model combined with the genetic algorithm sought to minimize the sum of the maximum stress and the stress distribution STD as the optimization objectives. The objective of achieving a uniform stress distribution was attained by optimizing the operating error parameters and adjusting the design of the cycloidal gear profile. This endeavor yielded optimal solutions manifested as [0.0641 mm, 0.0058 mm, 0.0584°] and [0.0645 mm, 0.0083 mm, 0.0819°], underscoring the promise and relevance of the approach and illuminating avenues for future advancements and potential applications in the realm of gear end face stress analysis.

**Author Contributions:** Conceptualization, N.J., S.W., and J.Z.; methodology, N.J. and L.J.; software, N.J. and L.J.; validation, N.J. and J.Z.; formal analysis, N.J.; investigation, N.J.; resources, N.J.; data curation, N.J.; writing—original draft preparation, N.J. and L.J.; writing—review and editing, N.J.; visualization, N.J. and L.J.; supervision, J.Z.; project administration, S.W.; funding acquisition, S.W. All authors have read and agreed to the published version of the manuscript.

**Funding:** This work was funded in part by the Hubei Province Technology Innovation Special Fund (No. 2019AAA069) and in part by the State Key Laboratory of New Textile Materials and Advanced Processing Technologies under Grant (no. FZ 20230023).

**Institutional Review Board Statement:** Not applicable.

**Data Availability Statement:** Not applicable.

**Conflicts of Interest:** The authors declare no conflict of interest.

## References

1. Hwang, S.-C.; Lee, J.-H.; Lee, D.-H.; Han, S.-H.; Lee, K.-H. Contact stress analysis for a pair of mating gears. *Math. Comput. Model.* **2013**, *57*, 40–49. [CrossRef]
2. Matejic, M.; Blagojevic, M.; Disic, A.; Matejic, M.; Milovanovic, V.; Miletic, I. A Dynamic Analysis of the Cycloid Disc Stress-Strain State. *Appl. Sci.* **2023**, *13*, 4390. [CrossRef]

3.  Li, S. Design and strength analysis methods of the trochoidal gear reducers. *Mech. Mach. Theory* **2014**, *81*, 140–154. [CrossRef]
4.  Liu, H.; Liu, H.; Zhu, C.; Tang, J. Study on gear contact fatigue failure competition mechanism considering tooth wear evolution. *Tribol. Int.* **2020**, *147*, 106277. [CrossRef]
5.  Liu, H.; Liu, H.; Bocher, P.; Zhu, C.; Sun, Z. Effects of case hardening properties on the contact fatigue of a wind turbine gear pair. *Int. J. Mech. Sci.* **2018**, *141*, 520–527. [CrossRef]
6.  He, H.; Liu, H.; Zhu, C.; Tang, J. Study on the gear fatigue behavior considering the effect of residual stress based on the continuum damage approach. *Eng. Fail. Anal.* **2019**, *104*, 531–544. [CrossRef]
7.  Guan, L.K.; Wang, N.N. Fracture Reason Analysis for Helical Gear Shaft of Converter Tilting Mechanism Senior Reducer. *Adv. Mater. Res.* **2012**, *605–607*, 824–827. [CrossRef]
8.  Qian, W.X.; Yin, X.W.; Xie, L.Y. Finite Element Analysis of a Certain Reducer Gear. *Adv. Mater. Res.* **2010**, *108–111*, 1066–1069. [CrossRef]
9.  Jalaja, K.; Manwatkar, S.K.; Gupta, R.K.; Narayana Murty, S.V.S. Metallurgical failure analysis of AA 7050 main landing gear brake reducer for reusable launch vehicle. *Eng. Fail. Anal.* **2023**, *145*, 107042. [CrossRef]
10. Yin, R.C.; Bradley, R.; Al-Meshari, A.; Al-Yami, B.S.; Al-Ghofaili, Y.M. Premature failure of a spiral bevel pinion in a turbine condenser gear reducer. *Eng. Fail. Anal.* **2006**, *13*, 727–731. [CrossRef]
11. Tsai, Y.T.; Lin, K.H. Dynamic Analysis and Reliability Evaluation for an Eccentric Speed Reducer Based on Fem. *J. Mech.* **2020**, *36*, 395–403. [CrossRef]
12. Li, X.; Shao, W.; Tang, J.; Ding, H.; Zhou, W. An Investigation of the Contact Fatigue Characteristics of an RV Reducer Crankshaft, Considering the Hardness Gradients and Initial Residual Stress. *Materials* **2022**, *15*, 7850. [CrossRef] [PubMed]
13. Feng, W.; Feng, Z.; Mao, L. Failure analysis of a secondary driving helical gear in transmission of electric vehicle. *Eng. Fail. Anal.* **2020**, *117*, 104934. [CrossRef]
14. Hou, X.; Gao, S.; Qiu, L.; Li, Z.; Zhu, R.; Lyu, S.-K. Transmission Efficiency Optimal Design of Spiral Bevel Gear Based on Hybrid PSOGSA (Particle Swarm Optimization—Gravitational Search Algorithm) Method. *Appl. Sci.* **2022**, *12*, 10140. [CrossRef]
15. Michaelis, K.; Seabra, J.; Höhn, B.R.; Hinterstoißer, M. Influence factors on gearbox power loss. *Ind. Lubr. Tribol.* **2011**, *63*, 46–55. [CrossRef]
16. Sun, X.; Han, L. Dynamics analysis of cbr reducer with tooth modification. In *Proceedings of the IOP Conference Series: Materials Science and Engineering*; IOP Publishing: Bristol, UK, 2019; p. 012007. [CrossRef]
17. Song, L.; Shunke, L.; Zheng, Z.; Chen, F. Analysis of the key structures of rv reducer based on finite element method. In *Proceedings of the IOP Conference Series: Materials Science and Engineering*; IOP Publishing: Bristol, UK, 2019; p. 012005. [CrossRef]
18. Hsieh, C.-F. Traditional versus improved designs for cycloidal speed reducers with a small tooth difference: The effect on dynamics. *Mech. Mach. Theory* **2015**, *86*, 15–35. [CrossRef]
19. Mo, J.; Luo, S.; Fu, S.; Chang, X. Meshing principle and characteristics analysis of an abnormal cycloidal internal gear transmission. *J. Mech. Sci. Technol.* **2022**, *36*, 5165–5179. [CrossRef]
20. Hsieh, C.-F. The effect on dynamics of using a new transmission design for eccentric speed reducers. *Mech. Mach. Theory* **2014**, *80*, 1–16. [CrossRef]
21. Li, X.; Li, C.; Wang, Y.; Chen, B.; Lim, T.C. Analysis of a Cycloid Speed Reducer Considering Tooth Profile Modification and Clearance-Fit Output Mechanism. *J. Mech. Des.* **2017**, *139*, 033303. [CrossRef]
22. Huang, C.; Wang, J.-X.; Xiao, K.; Li, M.; Li, J.-Y. Dynamic characteristics analysis and experimental research on a new type planetary gear apparatus with small tooth number difference. *J. Mech. Sci. Technol.* **2013**, *27*, 1233–1244. [CrossRef]
23. Bao, J.; He, W.; Qiao, S.; Johnson, P. Optimum design of parameters and contact analysis of cycloid drive. *J. Comput. Methods Sci. Eng.* **2021**, *21*, 71–83. [CrossRef]
24. Gao, P.; Yu, T.; Zhang, Y.; Wang, J.; Zhai, J. Vibration analysis and control technologies of hydraulic pipeline system in aircraft: A review. *Chin. J. Aeronaut.* **2021**, *34*, 83–114. [CrossRef]
25. Driscoll, T.A. Algorithm 756: A MATLAB toolbox for Schwarz-Christoffel mapping. *ACM Trans. Math. Softw.* **1996**, *22*, 168–186. [CrossRef]
26. Driscoll, T.A.; Trefethen, L.N. *Schwarz-Christoffel Mapping*; Cambridge University Press: Cambridge, UK, 2002; Volume 8.
27. Nasser, M.M. Fast computation of the circular map. *Comput. Methods Funct. Theory* **2015**, *15*, 187–223. [CrossRef]
28. Andreev, V.V.; McNicholl, T.H. Estimating the Error in the Koebe Construction. *Comput. Methods Funct. Theory* **2013**, *11*, 707–724. [CrossRef]
29. Jiang, N.; Wang, S.; Xie, X.; Yuan, X.; Yang, A.; Zhang, J. A vectorial modification methodology based on an efficient and accurate cycloid tooth profile model. *Precis. Eng.* **2022**, *73*, 435–456. [CrossRef]

*applied sciences*

*Article*

# A Dynamic Analysis of the Cycloid Disc Stress-Strain State

Milos Matejic [1], Mirko Blagojevic [1], Aleksandar Disic [1], Marija Matejic [2], Vladimir Milovanovic [1] and Ivan Miletic [1,*]

[1] Faculty of Engineering, University of Kragujevac, Sestre Janjic 6, 34000 Kragujevac, Serbia
[2] Faculty of Technical Sciences, University of Pristina with Temporary Settled in Kosovska Mitrovica, Knjaza Milosa 7, 38220 Kosovska Mitrovica, Kosovo
* Correspondence: imiletic@kg.ac.rs; Tel.: +381-64-670-55-67

**Featured Application: In this study, an original approach to the dynamic analysis of cycloid disc stress-strain state is proposed, which provides a very accurate picture of dynamic processes in cycloid drives.**

**Abstract:** The problem of internal forces that occur on the cycloid disc during the cycloid speed reducer operation so far has not been considered in a way that reflects its actual workloads and stresses in the cycloid disc itself. This paper presents a dynamic analysis of the stress-strain state of a cycloid disc by using experimental and numerical methods. The following cases of meshing are presented in the paper: a single-tooth, double-tooth, and triple-tooth meshing of the cycloid disc and the ring gear. The cycloid disc was chosen for this study because it is one of the main elements and the most critical element of the cycloid speed reducer. An experimental physical model of the cycloid disc and the meshing elements of the cycloid speed reducer was made based on a previously performed 3D CAD model. The numerical analysis of the stress-strain state of the cycloid disc was performed with the identically defined external load using the transient stress method. The paper presents a comparative analysis of the experimental and numerical results, which gives a solid insight into what is happening in the cycloid disc during the cycloid speed reducer operation. The experimental and simulation results both give the results with a deviation between 3% and 15%. After the detailed analyses, it is shown that the most critical element of cycloid speed reducer are output rollers, which need further study.

**Keywords:** cycloid disc; cycloid speed reducer; stress-strain state; experimental analysis; numerical analysis

**Citation:** Matejic, M.; Blagojevic, M.; Disic, A.; Matejic, M.; Milovanovic, V.; Miletic, I. A Dynamic Analysis of the Cycloid Disc Stress-Strain State. *Appl. Sci.* **2023**, *13*, 4390. https://doi.org/10.3390/app13074390

Academic Editor: Alberto Campagnolo

Received: 20 February 2023
Revised: 15 March 2023
Accepted: 18 March 2023
Published: 30 March 2023



## 1. Introduction

Over the last few decades, cycloid speed reducers have found a wide application in engineering practice. They are generally used as reducers, while their applications as multipliers or two-way power transmission systems have not been sufficiently studied yet. Cycloid speed reducers have a great number of good performance characteristics, such as a wide range of transmission ratios, extremely compact design, low noise and vibration, lightweight concerning the torque they transmit, ability to transmit high torques, reliable operation under intensive dynamic loads, high efficiency, etc. All these good operating characteristics together with a good price range, give them an advantage over conventional speed reducers. These gear trains have found application in robots, satellites, manipulation devices, crane machines, large industrial mixers, conveyors, and renewable energy sources (wind generators, mini hydro power plants).

The cycloid speed reducer force analysis and stresses in reducer elements caused by internal forces is a very complex task. For that reason, the cycloid speed reducer elements are often over-dimensioned, considering the possible problems that can be encountered in the cycloid speed reducer operation. The first analysis of the cycloid speed reducer internal forces was performed by Kudrijavcev in his book *Planetary Gear Train* [1]. His model was further improved by Lehmann [2] in his doctoral thesis. Malhorta and Parameswaran [3] developed a detailed model for determining the forces on all the elements of the cycloid speed reducer, which they later used to calculate the efficiency. The further published research about cycloid reducer dynamic analysis can be divided into two main groups: the papers that deal mainly with analytical models and simulations and the papers with dynamic simulations compared with experimental testing. One of the first papers from the first group of research was performed by Thube and Bobak [4], showing the detailed dynamic simulation in CAD software of a cycloid speed reducer without a theoretical background. On the other hand, Kniazeva and Goman [5] gave the analytical method of excitation force creation and its transformation into internal reducer forces. Furthermore, Wang et al. [6] analyzed sliding speed in cycloid speed reducers. Li [7] did a static analysis of the stress state in cycloid discs interacting with other reducer meshing elements. Hsieh [8,9] did both simulation and experimental analysis in classic cycloid speed reducers and non-pin wheel cycloid speed reducers, comparing their results. Shen et al. [10] did a dynamic analysis of cycloid speed reducers with special attention to their vibrations. Other authors in this field [11–17] did various dynamic simulations of cycloid speed reducer working using different software for its analysis, such as SolidWorks, ANSYS, MSC NASTRAN, etc. Two papers [18–20] in this field have been different because they described the simulation of contact stresses in cycloid discs all along their width. One of the most influential factors in the cycloid speed reducer dynamic is its machining and manufacturing tolerances [21]. Very accurate research about forces and stress calculation was performed by Efremenkov et al. [22], which can be used as a starting point for further experimental analyses.

Papers from the second group of papers related to cycloid reducer dynamics involve investigations with experimental testing. These papers were not published often because of the experiments' complexity and their costs. One of the first papers related to cycloid reducer dynamics was written by Hsieh [8], and the part related to the experiment was conducted in an open power circuit. In that research, the output torque was measured in two types of reducers: classic cycloid reducer and non-pin wheel concept. Another publication that had an experiment was by Kumar et al. [23]. After analysis in ANSYS, they found that the existing mechanism of the cycloid speed reducer was a critical part and they performed an experiment on the real model. The vibration analysis, numerically and experimentally, was conducted by Chen and Yang [24]. The same group of authors also performed one more experiment with a classic type of cycloid speed reducer and a new proposed model [25]. In that experiment, they used an accelerometer on the reducer output to have insight into the output shaft starting and stopping. They also conducted additional research with the same model, where they performed vibration analyses and oscillation modes [25]. Recent research in this field [26] shows the proposed design of a cycloid speed reducer with experimental verification of its functioning.

The authors of this research also have a background in this field, both theoretical and experimental papers. A solid foundation in the field of cycloid reducer research was conducted by Blagojevic [27] in his doctoral thesis. Furthermore, Blagojevic et al. [28] set up the single-stage cycloid reducer dynamic model, solved in MATLAB. In later research, a group of authors [29] analyzed the new concept of a two-stage cycloid reducer that uses one cycloid disc instead of two per transmission stage. This research uses an experiment of cycloid disc static loading. Blagojevic et al. [30] conducted FEM and experimental analyses on one cycloid disc from a cycloid speed reducer in static conditions. Interesting research by Blagojevic and a group of authors [31] was conducted similarly to [7], but NASTRAN was used for the solving software. For the new concept of a cycloid speed reducer presented in Blagojevic's Ph.D. thesis, a complete dynamic model has been given in the paper [32]. In another paper [33], the authors gave complete static analyses for a particular cycloid speed reducer with forces calculated on each reducer element. Blagojevic et al. [34] investigate the influence of friction forces on cycloid drive efficiency. In that paper, they show that friction forces can greatly impact cycloid drive operation due to decreased lubrication in case of oil leakage, lubricant decreased properties during operation, etc.

During dynamic load conditions, power transmissions must keep their stability, precision, and endurance. Therefore, examining cycloid speed reductions in dynamic load conditions is very significant. As can be noticed, in the available literature, most papers have analytical and numerical results. Few papers have an experiment. Most of the shown experiments in the available literature are conducted under static load conditions. The problem with static load conditions is that a physical process in the actual working conditions of the cycloidal drive is not simulated properly. Some of the dynamic experiments shown in the literature review only give output values of torque or output of acceleration. These experiments do not give complete insight into what is happening in the cycloid speed reducer during meshing in working conditions. Thus far, the experimental dynamic loading of the cycloid disc in laboratory conditions is not published because the examination methodology is very complex. In this paper, the authors will present a new approach to cycloid disc dynamic testing as well as cycloid disc mesh elements (output pin and meshing tooth). The testing was conducted both experimentally and numerically.

The rest of the paper is structured as follows: Section 2 presents the loading cases of the cycloid disc and its theoretical background. Section 3 presents experimental testing. In Section 4, the numerical testing was shown. Section 5 deals with the discussion of the results. At the end of the paper, in Section 6, conclusions are drawn.

## 2. Theoretical Model of the Cycloid Disc Loading

This paper studies a classic single-stage cycloid speed reducer with two cycloid discs shown in Figure 1. The paper is orientated in cycloid disc dynamic analyses and elements that interact with it during the cycloid speed reducer operation.

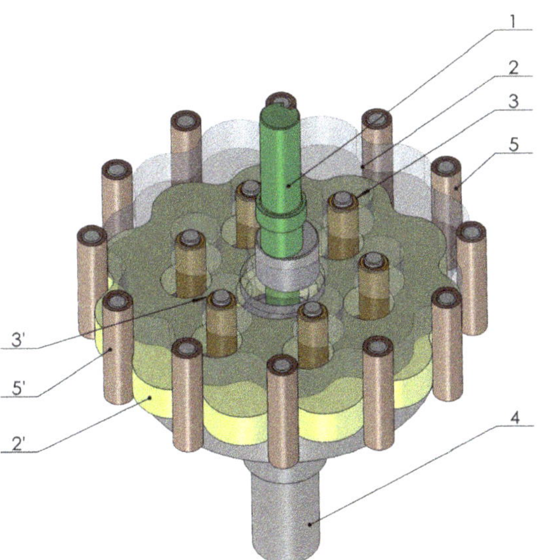


**Figure 1.** 3D model of a classic single-stage cycloid speed reducer: 1—input shaft; 2 and 2′—cycloid discs; 3 and 3′—output rollers; 4—exiting mechanism, and 5 (5′)—ring gear rollers.

A cycloid speed reducer is a very complex system from the aspect of its geometry as well as from the aspect of its kinematic and dynamic structure. In order to perform the analysis, it is necessary to know the loads that act on the cycloid disc as the main element of the cycloid speed reducer. Figure 2 shows the forces acting on the cycloid disc during meshing. They are the following [1,2]:

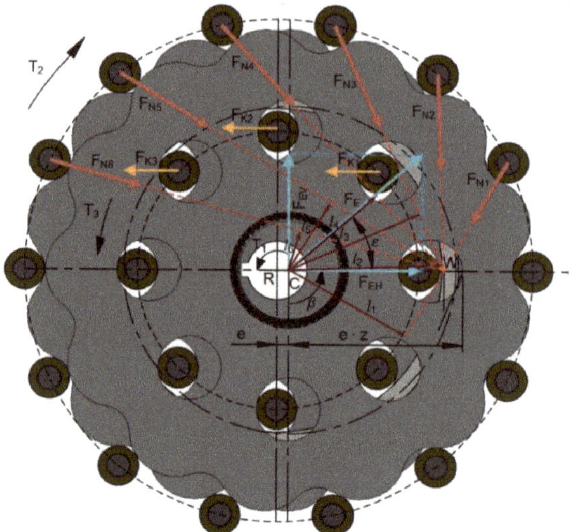

**Figure 2.** Forces on the cycloid disc [35].

-   $F_E$—eccentric cam force (the force where the vertical component $F_{EV}$ generates the drive torque $T_1$ on the cycloid disc due to its eccentric rotation),
-   $F_N$—normal force at the current contact point of the cycloid disc tooth and the stationary ring gear roller (normal force),
-   $F_K$—normal force at the current contact point of the output roller and the opening in the cycloid disc (output force).

The torques due to which these forces occur are:

-   $T_1$—drive torque of the cycloid disc,
-   $T_2$—torque on the ring gear,
-   $T_3$—output torque of the cycloid disc.

From Figure 2, it is evident that this is a statically indeterminate system. The only known force is the vertical component of the eccentric cam force $F_{EV}$. Kudrijavcev defined the first procedure for calculation of the forces acting on the cycloid disc in his book *Planetary Gear Train* in 1966 [1]. Both Kudrijavcev and Lehmann [2], as pioneers in this field, analyzed the ideal case of the cycloid disc meshing with other elements of the cycloid speed reducer. In this case, there are no clearances between the teeth of the cycloid disc and the rollers of the stationary ring gear, hence the assumption that half of the teeth of the cycloid disc transfer the load simultaneously. This is obviously not the case in real operating conditions. In reality, clearances exist for many reasons: as the result of manufacturing faults, to facilitate assembly and disassembly for maintenance purposes, and to enable adequate lubrication, among other reasons. Kudrijavcev's procedure for calculating the normal force occurring in the contact of the cycloid disc and the stationary ring gear roller is described in great detail in the literature [1]. In this paper, only the basic expressions are presented in order to help understand the calculation of the normal force in experimental and numerical analyses.

The normal force on the *i*-th roller is calculated based on Equation (1):

$$F_{Ni} = F_{Nmax} \cdot \frac{l_i}{r_1} \tag{1}$$

where $F_{Nmax}$—is the maximum value of the normal force, $l_i$—is the normal distance from the center of the cycloid disc to the corresponding normal force (Figure 2), $r_1$—is the radius of the stationary circle [1].

Since the output torque of the cycloid disc $T_3$ is given by Equation (2):

$$T_3 = F_{Nmax} \cdot r_1 \cdot z_2 \cdot \frac{\sum_i l_i^2}{r_1^2 \cdot z_2} \tag{2}$$

Having solved Equation (2), the maximum value of the normal force can be calculated as in Equation (3):

$$F_{Nmax} = \frac{4 \cdot T_3}{r_1 \cdot z_2} \tag{3}$$

where $z_2$ is the number of the rollers of the stationary ring gear.

The size of the clearances between the ring gear and the cycloid disc is directly related to the number of the ring gear rollers meshing with the cycloid disc. The larger the clearances, the smaller the number of the meshed teeth of the cycloid disc and the rollers of the ring gear. In this paper, three types of meshing were analyzed: triple-tooth, double-tooth, and single-tooth meshing. The single-tooth meshing is the most unfavorable type because only one tooth of the cycloid disc and one roller of the ring gear is meshing. The cycloid disc of the single-stage cycloid speed reducer (Figure 2) used in this paper has the following characteristics:

- Input power: $P_{in}$ = 5.5 kW;
- Input number of revolutions: $n_{in}$ = 1450 min$^{-1}$;
- Gear ratio: $u_r$ = 11;
- Eccentricity $e$ = 4 mm.

Numerical and experimental analyses were performed for all three types of meshing and for three different loads: 50% of $F_{Nmax}$, 100% of $F_{Nmax}$, and 150% of $F_{Nmax}$. The overload of 150% of $F_{Nmax}$ is taken into account in cases of shock loads or increased friction loads due to decreased lubrication occurring during the cycloid drive operation. Based on Equation (1), the normal forces were calculated as a function of the driving angle from the moment the cycloid disc teeth entered the mesh up to the moment they exited the mesh. Until the cycloid disc teeth enter the mesh, the normal meshing force equals zero. This procedure was carried out for all three types of meshing: single-tooth, double-tooth, and triple-tooth meshing. Based on the obtained values for the normal force, an approximate function of the normal force was obtained by function fitting and used in the experimental and numerical analyses. Approximate functions of the normal force are given in Table 1.

**Table 1.** Approximate functions of the normal force in the meshing of the cycloid disc teeth and the ring gear.

| Meshing Type | Approximate Function of the Normal Force for a Single Revolution of the Input Shaft | $F_{low}$, N | $F_{amp}$, N |
|---|---|---|---|
| Single tooth meshing | $F(\varphi) = \begin{cases} 0, & [0,20) \\ F_{low} + \lvert F_{amp} \cdot \sin(1.5 \cdot \varphi) \rvert, & [20,120] \\ 0, & (120,360] \end{cases}$ | 2387 | 1616.8 |
| Double tooth meshing | $F(\varphi) = \begin{cases} 0, & [0,20) \\ F_{low} + \lvert F_{amp} \cdot \sin(2 \cdot \varphi) \rvert, & [20,90] \\ 0, & (90,360] \end{cases}$ | 1530 | 427.7 |
| Triple tooth meshing | $F(\varphi) = \begin{cases} 0, & [0,30) \\ F_{low} + \lvert F_{amp} \cdot \sin(1.5 \cdot \varphi) \rvert, & [30,90] \\ 0, & (90,360] \end{cases}$ | 1456 | 293.9 |

Based on the obtained functions given in Table 1, a diagram of the normal force was prepared. The diagram of the normal force $F_n$ as a function of the driving angle $\varphi$ is shown in Figure 3.

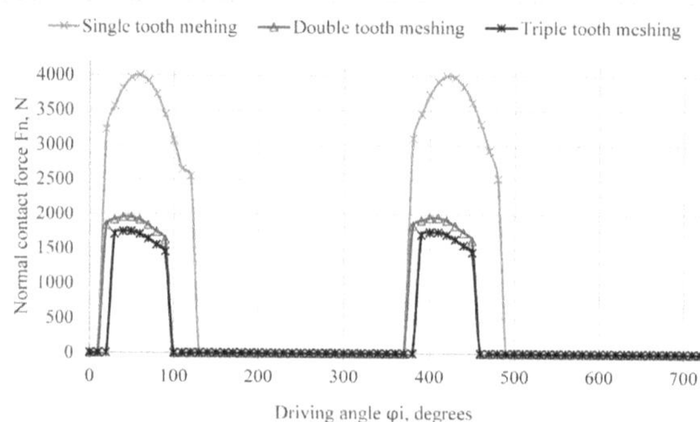



**Figure 3.** Diagram of the normal contact force.

## 3. Experimental Analysis of the Cycloid Disc

The experimental testing was performed at the Center for Engineering Software and Dynamic Testing of the Faculty of Engineering, the University of Kragujevac, on a Shimadzu EHF-EV101KZ-070-0A servo-hydraulic pulsator, with an accuracy of ±0.5% [35]. The test was performed at room temperature (22 ± 2 °C) by controlling the force according to the given input functions (Figure 3). Based on the input forces as a function of the driving angle ϕ (Figure 3), the function was transferred into a time domain corresponding to the rotational speed of the input shaft of the cycloid speed reducer of 350 min$^{-1}$. This was the maximum speed of loading and unloading that could be achieved on the servo-hydraulic pulsator. Higher speeds could not be achieved due to the inertia of the servo-hydraulic pulsator system. The diagram of the input force achieved on the servo-hydraulic pulsator is given in Figure 4.

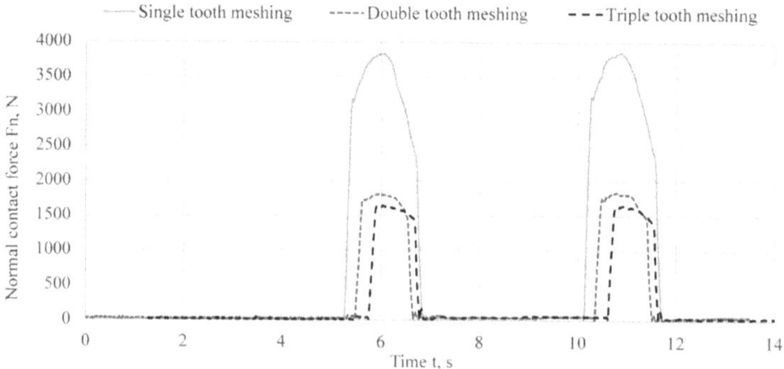



**Figure 4.** The input force achieved at the servo-hydraulic pulsator.

If Figures 3 and 4 are compared, it can be concluded that a precise application of force was achieved using the pulsator. The experimental model on the described pulsator is shown in Figure 5. The direction of eccentricity makes a 20° angle in the horizontal direction. At this position of the eccentric cam, all the elements come into contact, and the compression force assumes a vertical direction. This is a necessary condition for testing on a servo-hydraulic pulsator.

The strain gauge method was used for experimental analysis. Due to space limitations, the strain gauges could not be placed in the contact zone, so they were placed as close to it as possible. We used strain gauge chains 1-KY11-2/120 produced by HBM, one of the world's leading test and measuring equipment manufacturers, with an accuracy of

$\pm 1\%$ [36]. Two chains were glued. Each of the glued chains contained 5 strain gauges spaced at 2 mm, which made it possible to measure the strain gradient. The area of a single strain gauge was 1.95 mm$^2$. A schematic presentation of the position of the strain gauges is given in Figure 6.

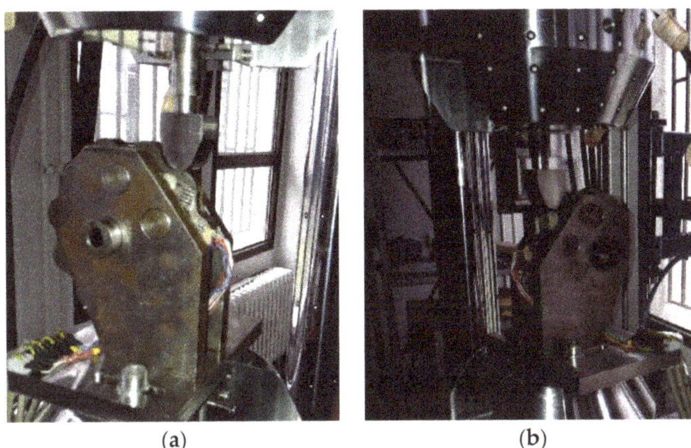

(**a**)　　　　　(**b**)

**Figure 5.** The experimental model on the servo-hydraulic pulsator: (**a**) left side view; (**b**) right side view 2.

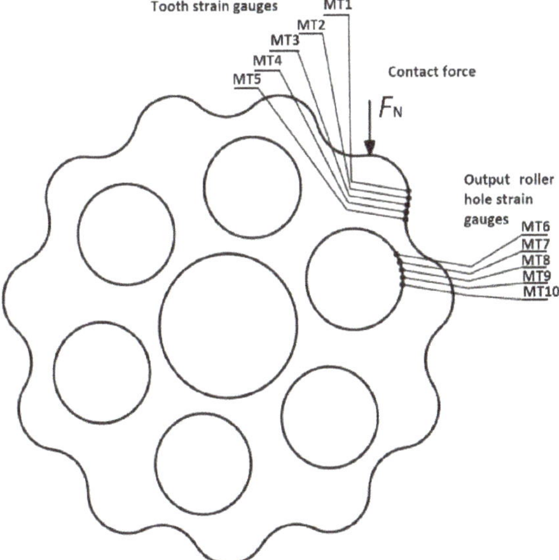


**Figure 6.** Position of the glued strain gauges.

The maximum values given in Table 2 were used to set the maximum compression force. A total of 9 measurements were made for all three meshing cases with load levels of 50%, 100%, and 150% of the force $F_{Nmax}$. The first measurement was made for the case of a triple tooth-meshing of the cycloid disc and a ring gear.

This paper presents strain measurement results for the maximum load of 150% of $F_{Nmax}$. These results are given in Figure 7.

**Table 2.** External loads during the meshing.

| Meshing | Load 50% $F_{Nmax}$, N | Load 100% $F_{Nmax}$, N | Load 150% $F_{Nmax}$, N |
|---|---|---|---|
| Triple tooth | 872.5 | 1745 | 2617.5 |
| Double tooth | 975.5 | 1951 | 2926.5 |
| Single tooth | 2000.5 | 4001 | 6001.5 |

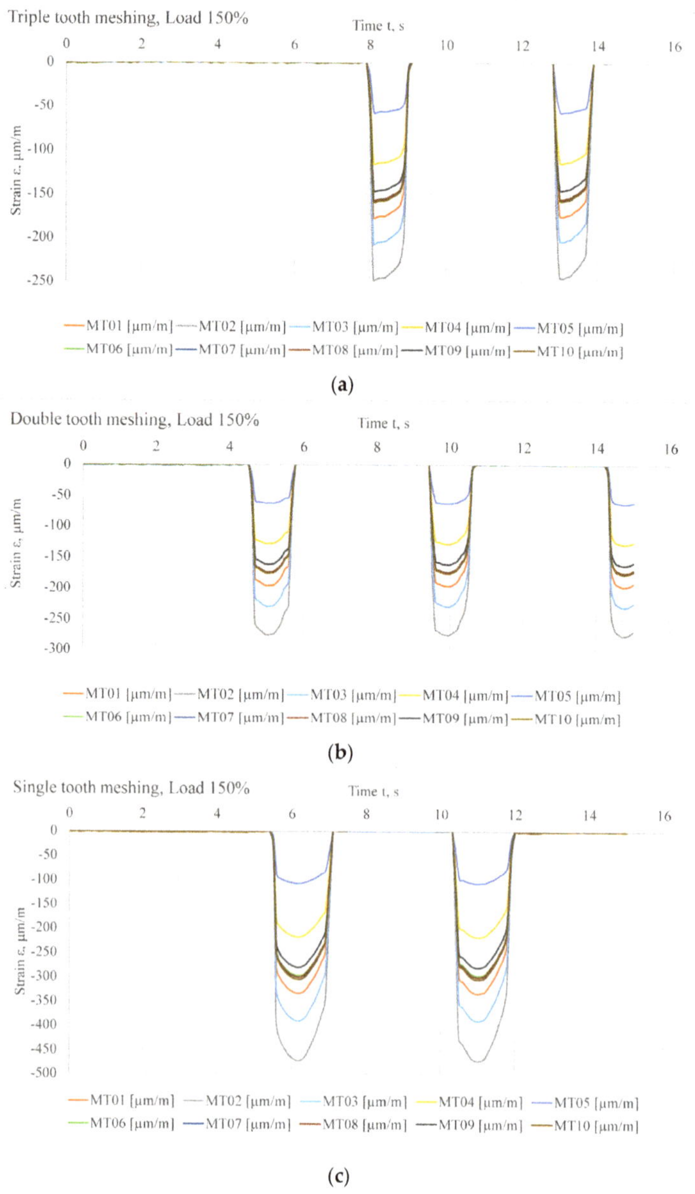

**Figure 7.** Strain results obtained in the time domain for the external load of 150% of $F_{nmax}$ for: (**a**) triple–tooth meshing, (**b**) double–tooth meshing, and (**c**) single–tooth meshing.

Figure 7, for the load case of 150% of $F_{Nmax}$, shows the results for three cases of meshing: triple tooth meshing, double tooth meshing, and single tooth meshing. The results present the measured strain by strain gauges. As can be noted, and as well expected, the most critical case is single tooth meshing because the biggest force was applied during that experiment.

## 4. Numerical Analysis of the Stress-Strain State of the Cycloid Disc

In order to verify the performed measurements, numerous numerical analyses were carried out using the finite element method. In both the experimental and numerical analyses for all three types of meshing, loading and unloading were performed to correspond to the actual going of the cycloid disc teeth and the ring gear into and out of the mesh. Namely, the tooth loading was gradually increased until the maximum force was applied, followed by a partial and then a complete tooth unloading until the moment the teeth re-entered the mesh. If Figures 5 and 8 are compared, it can be concluded that the experimental model is very similar to the CAD model used for the numerical analysis so that simulation can follow the experiment procedure.

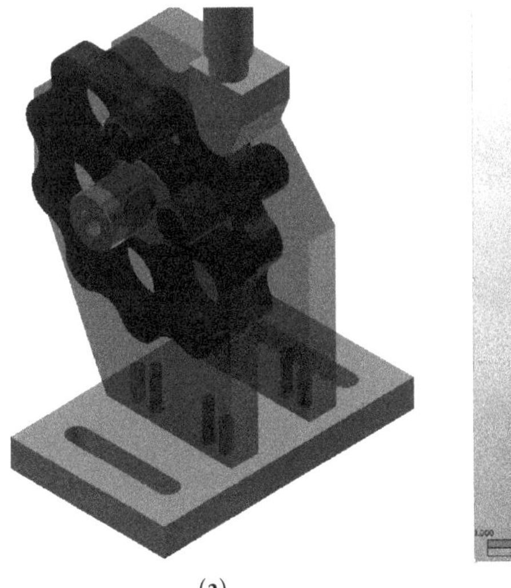

(a)

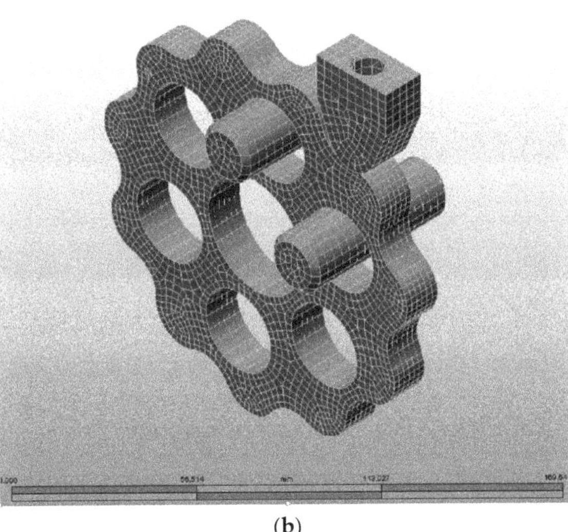

(b)

**Figure 8.** CAD model of a segment of the cycloid speed reducer: (**a**) experimental analysis (**b**) FEM analysis.

The numerical analysis of the stress-strain state of the cycloid disc was performed using the finite element method. The CAD model was made using the Autodesk Inventor software, while the numerical analysis was performed using the Autodesk Simulation Mechanical software, which uses the ALGOR solver. In order to be able to compare the numerical and experimental results, the CAD model for numerical analysis was made based on the physical model of the cycloid disc and other meshing elements used for experimental analysis (Figure 5). For the numerical analysis to be efficient, only the parts in direct contact were considered for the FEA model, while the rest were taken as constraints, Figure 8b.

The numerical analysis was performed as a dynamic analysis in the transient stress environment. Therefore, as part of this dynamic analysis, simulations were performed for a single-tooth, double-tooth, and triple-tooth meshing of the ring gear rollers and cycloid disc teeth. The function of the $F_{Nmax}$ was modeled in the real-time domain where the loading and unloading correspond to the actual going of the cycloid disc teeth and the ring gear rollers into and out of the mesh [1,2,30]. The cycloid disc was considered a deformable elastic body in the analyses.

Interactions between the meshed elements were modeled as surface contacts. Unlike in previous studies [30], where the line load was applied in the contact area with the pusher (which simulated the ring gear), the surface contact load transfer was used in this study. As illustrated in Figure 9, the pusher can move only in the vertical direction. In this way, a simulation that most closely represents the operating conditions of the cycloid disc was created. Based on the procedures described in the previous section and in the literature [1,2,30], the force $F_{Nmax}$ was determined for the triple-tooth, double-tooth, and single-tooth meshing. For all three types of meshing, the following external loads were used: 50% of $F_{Nmax}$, 100% of $F_{Nmax}$, and 150% of $F_{Nmax}$. Because of the large number of results obtained, only the results for the dynamic force with the maximum value of 150% of $F_{Nmax}$ modeled for single-tooth, double-tooth, and triple-tooth meshing were presented in the paper.

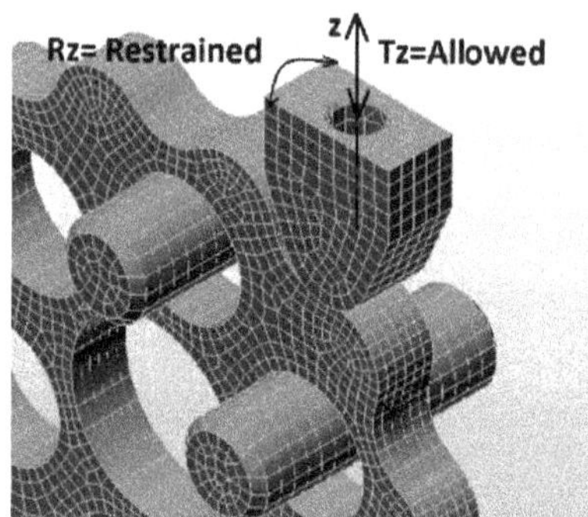


**Figure 9.** Illustration of the limitations of the pusher.

The values of the external loads used in the numerical analysis are given in Table 2. They correspond to the diagram of the forces achieved on the servo-hydraulic pulsator. Table 3 presents the ordinal numbers of the nodes that best correspond to the centers of the strain gauges (Figures 5 and 6). The von Mises stress is chosen for the comparison because it has a similar value to the dominant stress component. The deviation between the dominant stress component and von Mises stress is negligible, which is the main reason for choosing von Misses stress for the comparison of results.

**Table 3.** Comparative presentation of the strain gauges and the corresponding finite element nodes.

| Node Number | MT Number | Node Number | MT Number |
|---|---|---|---|
| Strain von Misses (1377) | MT1 | Strain von Misses (1846) | MT6 |
| Strain von Misses (1378) | MT2 | Strain von Misses (1847) | MT7 |
| Strain von Misses (1379) | MT3 | Strain von Misses (1848) | MT8 |
| Strain von Misses (1380) | MT4 | Strain von Misses (1849) | MT9 |
| Strain von Misses (1381) | MT5 | Strain von Misses (1850) | MT10 |

This problem was regarded as spatial so that the numerical and experimental analyses could be performed with the same boundary conditions. In order to compare the numerical and the experimental results, the strain values were given as a function of time for the positions where the strain gauges were glued in the experimental testing. The positions of the strain gauges are shown in Figure 6—position of the glued strain gauges.

The strain gauges were positioned identically to the ones shown in Figure 6. This was achieved by placing virtual sensors on the nodes of the corresponding finite elements.

Figure 10 shows the numerical results for all three cases of meshing under the load of 150% of $F_{nmax}$ (Table 2).

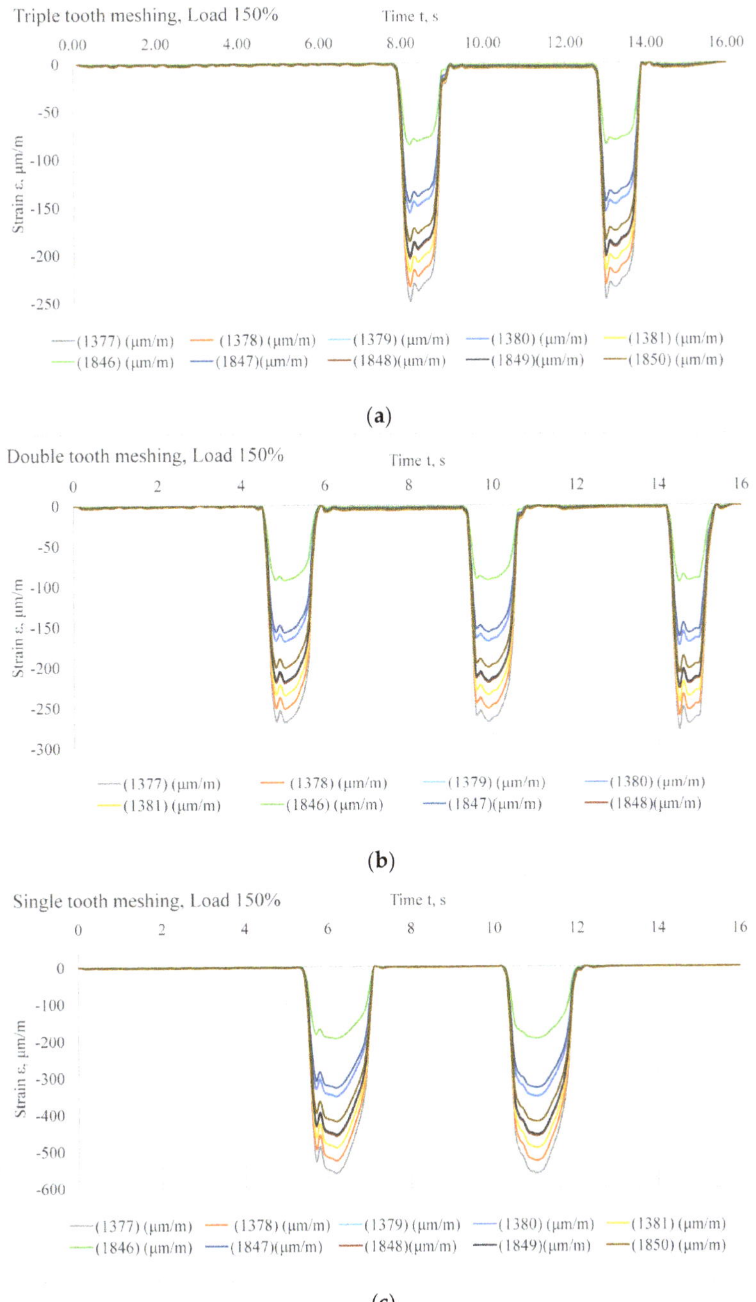

(**a**)

(**b**)

(**c**)

**Figure 10.** Von Mises strains in the time domain for the external load of 150% $F_{Nmax}$ for: (**a**) triple–tooth meshing, (**b**) double–tooth meshing, and (**c**) single–tooth meshing.

The cycloid disc was made of steel 30CrMoV9, which for the given thickness, according to the standard EN 10250-3:2000 [37], has a minimum yield stress of $R_e = 700$ MPa and a minimum tensile strength of $R_m = 900$ MPa. The other mechanical characteristics of the material needed for the numerical analysis were found in the software library: modulus of elasticity $E = 2.07 \times 10^5$ MPa and Poisson's ratio $\mu = 0.3$.

The lowest stress value on the cycloid disc occurs in the case of the triple-tooth meshing of the cycloid disc and the ring gear ($\sigma = 338$ MPa). About 7% higher stresses occur at the double-tooth meshing compared to the triple-tooth meshing, while about 224% higher stresses occur at the single-tooth meshing compared to the triple-tooth meshing. At the single-tooth meshing and the load of 150% of $F_{Nmax}$, the stresses that occur are close to the yield stress value, but this is only for very short time intervals.

According to the literature [23], the stress that occurs in the contact of the cycloid disc and the output rollers can have the most unfavorable effect on the operation of the cycloid speed reducer.

## 5. Experimental and Simulation Results Comparison

A comparative analysis of the numerical and experimental results is presented in this section. Table 4 gives the strain and stress values obtained using the finite element method and experimental testing. The values are given for the force that equals 150% of $F_{Nmax}$ for the most critical moment of loading.

**Table 4.** Comparative results obtained using the finite element method and experimental testing.

| Single-tooth meshing—Load 150% | | | | | | | | | | |
|---|---|---|---|---|---|---|---|---|---|---|
| | MT1 | MT2 | MT3 | MT4 | MT5 | MT6 | MT7 | MT8 | MT9 | MT10 |
| $\varepsilon$—ex., $\mu$m m$^{-1}$ | 473.75 | 390.26 | 333.91 | 218.09 | 106.51 | 297.37 | 300.36 | 300.63 | 280.83 | 305.38 |
| $\sigma$—ex., MPa | 98.07 | 80.78 | 69.12 | 45.14 | 22.05 | 61.56 | 62.18 | 62.23 | 58.13 | 63.21 |
| $\varepsilon$—num., $\mu$m m$^{-1}$ | 561.52 | 526.12 | 491.68 | 352.50 | 350.36 | 193.95 | 328.17 | 459.80 | 455.71 | 420.51 |
| $\sigma$—num., MPa | 116.24 | 108.91 | 101.78 | 72.97 | 72.52 | 40.15 | 67.93 | 95.18 | 94.33 | 87.04 |
| Double-tooth meshing—Load 150% | | | | | | | | | | |
| | MT1 | MT2 | MT3 | MT4 | MT5 | MT6 | MT7 | MT8 | MT9 | MT10 |
| $\varepsilon$—ex., $\mu$m m$^{-1}$ | 276.69 | 230.16 | 196.74 | 128.87 | 63.27 | 174.69 | 175.92 | 174.99 | 162.58 | 176.74 |
| $\sigma$—ex., MPa | 57.28 | 47.64 | 40.73 | 26.68 | 13.10 | 36.16 | 36.42 | 36.22 | 33.65 | 36.59 |
| $\varepsilon$—num., $\mu$m m$^{-1}$ | 277.21 | 259.73 | 242.73 | 174.02 | 172.96 | 95.75 | 162.01 | 226.99 | 224.97 | 207.59 |
| $\sigma$—num., MPa | 57.38 | 53.76 | 50.25 | 36.02 | 35.80 | 19.82 | 33.54 | 46.99 | 46.57 | 42.97 |
| Triple-tooth meshing—Load 150% | | | | | | | | | | |
| | MT1 | MT2 | MT3 | MT4 | MT5 | MT6 | MT7 | MT8 | MT9 | MT10 |
| $\varepsilon$—ex., $\mu$m m$^{-1}$ | 247.28 | 206.24 | 176.15 | 114.69 | 56.69 | 156.37 | 156.53 | 156.46 | 145.70 | 157.98 |
| $\sigma$—ex., MPa | 51.19 | 42.69 | 36.46 | 23.74 | 11.73 | 32.37 | 32.40 | 32.39 | 30.16 | 32.70 |
| $\varepsilon$—num., $\mu$m m$^{-1}$ | 249.47 | 233.74 | 218.45 | 156.61 | 155.66 | 86.17 | 145.80 | 204.28 | 202.47 | 186.82 |
| $\sigma$—num., MPa | 51.64 | 48.39 | 45.22 | 32.42 | 32.22 | 17.84 | 30.18 | 42.29 | 41.91 | 38.67 |

In both numerical and experimental analyses, the highest strain occurs at the site of strain gauge 1, i.e., at the strain von Mises node (1377). Figure 11 shows comparative strain diagrams of strain gauge 1 (MT1) and the corresponding strain von Mises node (1377) for the most unfavorable load when the force $F_N$ reaches 150%.

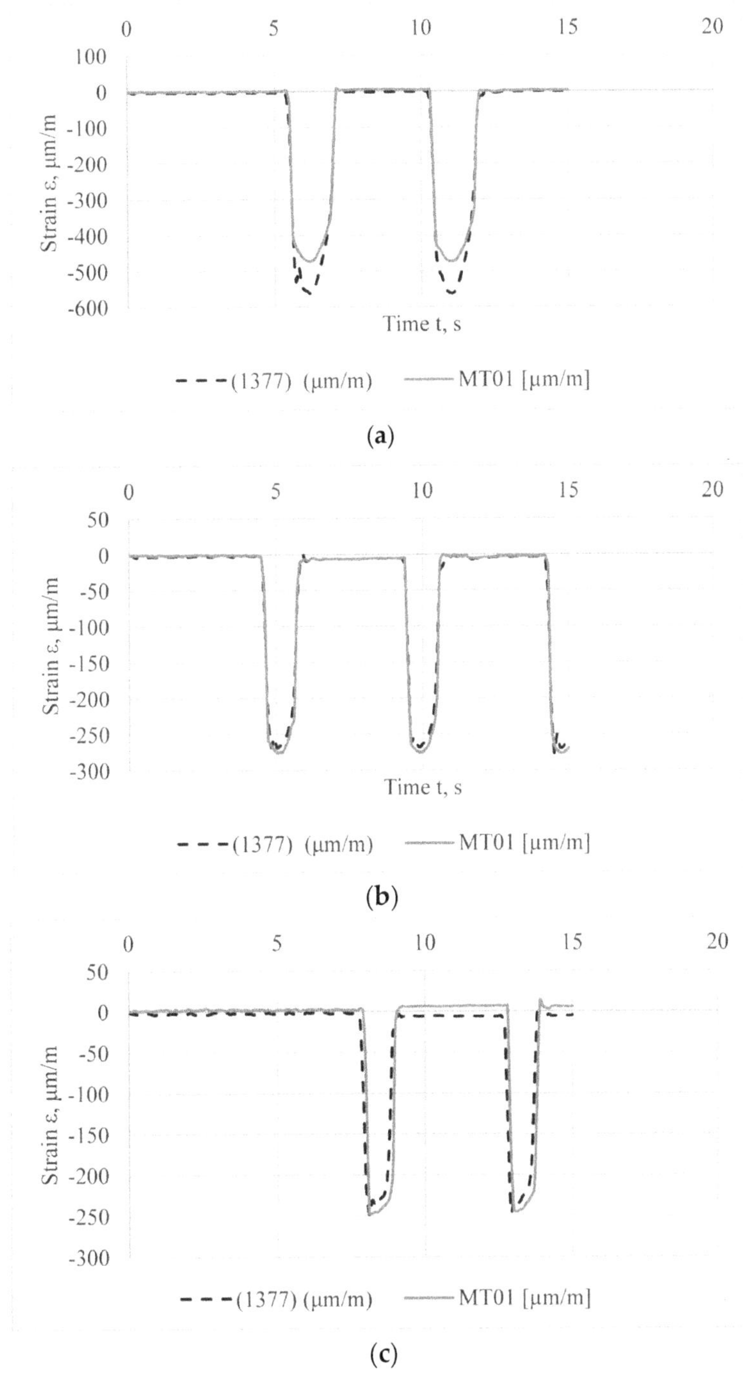



**Figure 11.** A comparative diagram of the numerical and experimental results obtained at site MT1 (node 1377) at the load of 150% of $F_N$ for (**a**) single–tooth meshing; (**b**) double–tooth meshing, and (**c**) triple–tooth meshing.

The stresses that occur in the case of the single-tooth meshing are almost twice as high as that of the double-tooth or triple-tooth meshing of the cycloid disc and the ring gear. This is found in both numerical and experimental analyses. The difference in stresses between the double-tooth and the triple-tooth meshing is much smaller than between the single-tooth and the double-tooth meshing. A correlation analysis was conducted for the presented diagrams. The single tooth meshing results correlates 99.6% between the experimental and numerical results. The double-tooth and triple-tooth meshing results correlate 99.7% and 92.9% between the experimental and numerical results. The biggest deviation in correlation was the triple tooth meshing case because of the multiple meshing contacts and harder input force prediction. The correlation analyses show that the simulation setup follows the experimental procedure in more than acceptable boundaries. The highest stresses occur in the simulations and experiments at the site of strain gauge 1 or node 1377. Figure 11 gives a comparative diagram of the strains at the site of strain gauge 1. Based on Figure 11, it can be concluded that the difference in the strain values is almost negligible in the double-tooth and triple-tooth meshing, while the strain values for the single-tooth meshing differ considerably. Namely, in the case of the triple meshing, the strain values obtained by numerical analysis are slightly lower compared to the values obtained experimentally. The lowest strains occur at the site of strain gauge 5 in the experimental analysis, while in the numerical analysis, the lowest strains occur at strain gauge 6. In the numerical analysis, the strain values are much higher at the sites of strain gauges 4 and 5, while at the site of strain gauge 6, the stress values are lower. This is not the case with other strain gauges. This inconsistency may be caused by improperly glued strain gauges or inhomogeneity of the cycloid disc material, among other reasons. The deviations between the numerical and experimental analysis results for the strain gauge 1 range from 3% to about 15%. Therefore, it can be concluded that the numerical results are valid and that the numerical model was well developed. The deviation between results probably occurred because the von Mises stress is considered instead of the stress component. The results deviation is at appropriate values.

## 6. Conclusions

This paper presents a numerical and experimental dynamic analysis of the stress state of the cycloid disc meshing with other elements of the cycloid speed reducer. A single-tooth, double-tooth, and triple-tooth meshing were analyzed. Based on the obtained results, the following can be concluded:

- There is a high level of matching between the numerical and the experimental results. The level of mismatching for the strain gauge with the largest measured strain, MT1 (node 1377), is between 3% and 15%. This clearly shows that the boundary conditions set for the numerical model corresponded to the operating conditions in the experiment. The presented deviation between the results is in the acceptable interval.
- As expected, the maximum stress occurs in the meshing zone of the cycloid disc and the ring gear. These stress values are far below the material yield stress.
- There are two more places where the stress spikes occur: in the contact area between the cycloid disc and the eccentric cam and in the contact area between the cycloid disc and the output rollers, as shown in [23]. These are expected results, as shown in the literature review.
- The stress values are far below the yield stress value of the cycloid disc material at the double-tooth and triple-tooth meshing, while in the case of the single-tooth meshing, the stress values are close to the yield stress value (the most unfavorable case being when the load was up to 150% of the force $F_{Nmax}$). Overload is included in this research for the cases of shock forces or increased friction forces occurring during the cycloid drive operation to cover those possibilities as well.
- Based on the performed analysis, it can be concluded that the output rollers of the cycloid speed reducer need to be further studied in more detail since the contact area between the output rollers and the cycloid disc is one of the places where the

highest stresses occur. This is the place that requires further analyses and possible new experimental setups.

- Manufacturing tolerances for cycloid speed reducers should be kept as low as possible, as shown in [21] so that at least a double-tooth meshing can be achieved. This would considerably decrease the stresses occurring during interactions between the elements of the cycloid speed reducer.

In further research, particular attention will be paid to other elements of the cycloid speed reducer, such as the eccentric cam, output rollers, and ring gear.

**Author Contributions:** Conceptualization, M.B. and M.M. (Milos Matejic); methodology, M.M. (Marija Matejic); software, V.M.; validation, V.M. and A.D.; formal analysis, M.M. (Marija Matejic); investigation, A.D.; resources, M.M. (Milos Matejic); data curation, A.D.; writing—original draft preparation, M.M. (Milos Matejic); writing—review and editing, M.B.; visualization, I.M.; supervision, V.M.; project administration, M.B.; funding acquisition, I.M. All authors have read and agreed to the published version of the manuscript.

**Funding:** This paper is funded through the EIT's HEI Initiative SMART-2M project, supported by EIT RawMaterials, funded by the European Union.

**Institutional Review Board Statement:** Not applicable.

**Informed Consent Statement:** Informed consent was obtained from all subjects involved in the study.

**Data Availability Statement:** Not applicable.

**Conflicts of Interest:** The authors declare no conflict of interest.

# References

1. Kudrijavcev, V.N. *Planetary Gear Train*; Leningrad: Moscow, Russia, 1966. (In Russian)
2. Lehmann, M. *Calculation and Measurement of Forces Acting on Cycloid Speed Reducer*; Technical University Munich: Munich, Germany, 1976. (In German)
3. Malhotra, S.K.; Parameswaran, M.A. Analysis of a cycloid speed reducer. *Mech. Mach. Theory* **1983**, *18*, 491–499. [CrossRef]
4. Thube, S.V.; Bobak, T.R. *Dynamic Analysis of a Cycloidal Gearbox Using Finite Element Method*; American Gear Manufacturers Association (AGMA): Alexandria, VA, USA, 2012; pp. 241–253.
5. Kniazeva, A.; Goman, A. Definition of a loading zone of a planetary pin reducer eccentric. *Meh. Mashin Meh. I Mater.* **2012**, *3*, 40–52.
6. Wang, S.; Tian, G.; Jiang, X. Estimation of sliding loss in a cycloid gear pair. *Int. J. Adv. Comput. Technol.* **2012**, *4*, 462–469. [CrossRef]
7. Li, S. Design and strength analysis methods of the trochoidal gear reducers. *Mech. Mach. Theory* **2014**, *81*, 140–154. [CrossRef]
8. Hsieh, C.F. Dynamics analysis of cycloidal speed reducers with pinwheel and nonpinwheel designs. *J. Mech. Des. Trans. ASME* **2014**, *136*, 091008. [CrossRef]
9. Hsieh, C.F. The effect on dynamics of using a new transmission design for eccentric speed reducers. *Mech. Mach. Theory* **2014**, *80*, 1–16. [CrossRef]
10. Zhang, L.; Wang, Y.; Wu, K.; Sheng, R.; Huang, Q. Dynamic modeling and vibration characteristics of a two-stage closed-form planetary gear train. *Mech. Mach. Theory* **2016**, *97*, 12–28. [CrossRef]
11. Zhang, F.; Li, P.; Zhu, P.; Yang, X.; Jiang, W. Analysis on dynamic transmission accuracy for RV reducer. *MATEC Web Conf.* **2017**, *100*, 01003. [CrossRef]
12. Ren, Z.Y.; Mao, S.M.; Guo, W.C.; Guo, Z. Tooth modification and dynamic performance of the cycloidal drive. *Mech. Syst. Signal Process.* **2017**, *85*, 857–866. [CrossRef]
13. Song, L.; Shunke, L.; Zheng, Z.; Chen, F. Analysis of the key structures of rv reducer based on finite element method. *IOP Conf. Ser. Mater. Sci. Eng.* **2019**, *544*, 012005. [CrossRef]
14. Bao, H.Y.; Jin, G.H.; Lu, F.X. Nonlinear dynamic analysis of an external gear system with meshing beyond pitch point. *J. Mech. Sci. Technol.* **2020**, *34*, 4951–4963. [CrossRef]
15. Król, R. Resonance phenomenon in the single stage cycloidal gearbox. Analysis of vibrations at the output shaft as a function of the external sleeves stiffness. *Arch. Mech. Eng.* **2021**, *68*, 303–320. [CrossRef]
16. Tsai, Y.T.; Lin, K.H. Dynamic Analysis and Reliability Evaluation for an Eccentric Speed Reducer Based on Fem. *J. Mech.* **2020**, *36*, 395–403. [CrossRef]
17. Tchufistov, E.A.; Tchufistov, O.E. Loading of satellite bearing of planetary cycloid gear by forces acting in meshing. *IOP Conf. Ser. Mater. Sci. Eng.* **2020**, *862*, 032042. [CrossRef]

18. Efremenkov, E.A.; Efremenkova, S.K.; Dyussebayev, I.M. Determination of Geometric Parameter of Cycloidal Transmission from Contact Strength Condition for Design of Heavy Loading Mechanisms. *IOP Conf. Ser. Mater. Sci. Eng.* **2020**, *795*, 012024. [CrossRef]
19. Zhang, T.; Li, X.; Wang, Y.; Sun, L. A semi-analytical load distribution model for cycloid drives with tooth profile and longitudinal modifications. *Appl. Sci.* **2020**, *10*, 4859. [CrossRef]
20. Liu, C.; Shi, W.; Xu, L. A Novel Approach to Calculating the Transmission Accuracy of a Cycloid-Pin Gear Pair Based on Error Tooth Surfaces. *Appl. Sci.* **2021**, *11*, 8671. [CrossRef]
21. Jiang, N.; Wang, S.; Yang, A.; Zhou, W.; Zhang, J. Transmission Efficiency of Cycloid–Pinion System Considering the Assembly Dimensional Chain. *Appl. Sci.* **2022**, *12*, 11917. [CrossRef]
22. Efremenkov, E.A.; Shanin, S.A.; Martyushev, N.V. Development of an Algorithm for Computing the Force and Stress Parameters of a Cycloid Reducer. *Mathematics* **2023**, *11*, 993. [CrossRef]
23. Kumar, N.; Kosse, V.; Oloyede, A. A new method to estimate effective elastic torsional compliance of single-stage Cycloidal. *Mech. Mach. Theory* **2016**, *105*, 185–198. [CrossRef]
24. Chen, C.; Yang, Y. Structural Characteristics of Rotate Vector Reducer Free Vibration. *J. Vibroeng.* **2016**, *18*, 3089–3103. [CrossRef]
25. Yang, Y.H.; Chen, C.; Wang, S.Y. Response sensitivity to design parameters of RV reducer. *Chin. J. Mech. Eng.* **2018**, *31*, 49. [CrossRef]
26. Lin, T.C.; Schabacker, M.; Ho, Y.L.; Kuo, T.C.; Tsay, D.M. Geometric Design and Dynamic Analysis of a Compact Cam Reducer. *Machines* **2022**, *10*, 955. [CrossRef]
27. Blagojević, M. Stress and Strain State of Cycloreducer's Elements under Dynamic Loads. Ph.D. Thesis, University of Kragujevac, Faculty of Mechanical Engineering, Kragujevac, Serbia, 2008.
28. Blagojevic, M.; Nikolic-Stanojevic, V.; Marjanovic, N.; Veljovic, L. Analysis of Cycloid Drive Dynamic Behavior. *Sci. Tech. Rev.* **2009**, *59*, 52–56.
29. Blagojevic, M.; Marjanovic, N.; Djordjevic, Z.; Stojanovic, B.; Disic, A. A new design of a two-stage cycloidal speed reducer. *J. Mech. Des. Trans. ASME* **2011**, *133*, 085001. [CrossRef]
30. Blagojevic, M.; Marjanovic, N.; Djordjevic, Z.; Stojanovic, B.; Marjanovic, V.; Vujanac, R.; Disic, A. Numerical and experimental analysis of the cycloid disc stress state. *Teh. Vjesn.* **2014**, *21*, 377–382.
31. Blagojevic, M.; Matejic, M. Stress and strain state of cycloid gear under dynamic loads. *Mach. Des.* **2016**, *8*, 129–132.
32. Blagojević, M.; Matejić, M.; Kostić, N. Dynamic behaviour of a two-stage cycloidal speed reducer of a new design concept. *Teh. Vjesn.* **2018**, *25*, 291–298. [CrossRef]
33. Blagojevic, M.; Matejic, M.; Vasic, M. Comparative Overview of Calculation of Normal Force on Cycloidal Gear Tooth. In Proceedings of the International Congress Motor Vehicles & Motors—Conference Porceedings, Kragujevac, Serbia, 9 October 2020; University of Kragujevac, Faculty of Engineering: Kragujevac, Serbia, 2020; pp. 131–137.
34. Blagojevic, M.; Kocic, M.; Marjanovic, N.; Stojanovic, B.; Dordevic, Z.; Ivanov; Marjanovic, V. Influence of the Friction on the Cycloidal Speed Reducer Efficeicncy. *J. Balk. Tribol. Assoc.* **2012**, *18*, 217–227.
35. SHIMADZU Dynamic and Fatigue Testing Systems Your Partner for Dynamic and Fatigue Tests. Shimadzu Cat. 2021. Available online: https://www.ssi.shimadzu.com/products/materials-testing/fatigue-testing-impact-testing/index.html (accessed on 20 February 2023).
36. HBM Strain Gauges: Absolute Precision from HBM. Straing Gauges Accessories Cat. 2014. Available online: https://www.straingauges.cl/images/blog/catalogo%20de%20strain%20gages.pdf (accessed on 15 January 2023).
37. *EN 10250-3:2000*; Open Steel Die Forgings for General Engineering Purposes Alloy Special Steels. British Standards Institution: London, UK, 1999.

 *applied sciences*

*Article*

# Multi-Objective Optimization of a Two-Stage Helical Gearbox Using Taguchi Method and Grey Relational Analysis

Xuan-Hung Le and Ngoc-Pi Vu *

Faculty of Mechanical Engineering, Thai Nguyen University of Technology, 3/2 Street, Tich Luong Ward, Thai Nguyen City 251750, Vietnam; lexuanhung@tnut.edu.vn
* Correspondence: vungocpi@tnut.edu.vn; Tel.: +84-974905578

**Abstract:** This paper presents a novel approach to solve the multi-objective optimization problem designing a two-stage helical gearbox by applying the Taguchi method and the grey relation analysis (GRA). The objective of the study is to identify the optimal main design factors that maximize the gearbox efficiency and minimize the gearbox mass. To achieve that, five main design factors, including the coefficients of wheel face width (CWFW) of the first and the second stages, the allowable contact stresses (ACS) of the first and the second stages, and the gear ratio of the first stage were chosen. Additionally, two single objectives, including the maximum gearbox efficiency and minimum gearbox mass, were analyzed. In addition, the multi-objective optimization problem is solved through two phases: Phase 1 solves the single-objective optimization problem in order to close the gap between variable levels, and phase 2 solves the multi-objective optimization problem to determine the optimal main design factors. From the results of the study, optimum values of five main design parameters for designing a two-stage helical gearbox were first introduced.

**Keywords:** gearbox; multi-objective optimization; gear ratio; two-stage helical gearbox

## 1. Introduction

Optimizing gearboxes is a critical aspect of mechanical engineering as it has a direct impact on their efficiency, durability, and reliability, as well as the performance of other machinery and equipment. Multi-objective optimization of gearboxes involves simultaneously optimizing various performance parameters, such as load-carrying capacity, noise, mass, size, and efficiency, making it a complex and challenging task. To address these challenges, various optimization techniques have been developed in recent years. Helical gearboxes, one of many gearbox types, are widely used in industrial applications due to their superior load-carrying capacity, smooth operation [1], simple structure, and low cost. However, designing a helical gearbox involves numerous design parameters, making it difficult to optimize for multiple objectives.

Numerous studies on the optimal design of helical gearboxes have been conducted thus far. This research looked into both single-objective and multi-objective optimization problems. Many authors are also interested in the single-objective optimization problem. I. Römhild and H. Linke [2] presented formulas for calculating gear ratios for two-, three-, and four-stage helical gearboxes in order to obtain the smallest gear mass. Milou et al. [3] presented a practical approach for reducing the mass of a two-stage helical gearbox in their paper. The method entailed analyzing data from gearbox manufacturers. Their findings suggested a center distance ratio $(aw_2/aw_1)$ between the second and first stages in the range of 1.4 to 1.6 to achieve the minimum mass. Once the optimal center distance ratio has been determined, the corresponding partial gear ratios are obtained using a lookup table. Various objective functions are utilized to solve the single-objective optimization problem of determining the optimal gear ratio of helical gearboxes. These objective functions include minimizing gearbox length [4–7], minimizing gearbox cross section area [6,8–10],

**Citation:** Le, X.-H.; Vu, N.-P. Multi-Objective Optimization of a Two-Stage Helical Gearbox Using Taguchi Method and Grey Relational Analysis. *Appl. Sci.* **2023**, *13*, 7601. https://doi.org/10.3390/app13137601

Academic Editor: Filippo Berto

Received: 24 May 2023
Revised: 12 June 2023
Accepted: 26 June 2023
Published: 27 June 2023

minimizing gearbox mass [6,11], minimizing gearbox volume [12], and minimizing gearbox cost [13–16]. All of these studies share a common feature in that only one design parameter, the partial gear ratio, is defined in the form of an explicit model.

The multi-objective optimization problem for designing helical gearboxes has also been of interest to researchers. M. Patil et al. [1] conducted a study in which a two-stage helical gearbox was subjected to multi-objective optimization with a broad range of constraints using a specially formulated discrete version of non-dominated sorting genetical algorithm II (NSGA-II). The study formulated two objective functions, namely the minimum gearbox volume and minimum total gearbox power loss. Moreover, the study considered constraints, such as bending stress, pitting stress, and tribological factors. Wu, Y.-R. and V.-T. Tran [17] introduced a new microgeometry modification for helical gear pairs, leading to substantial enhancements in performance with regards to noise and vibration. In their research, C. Gologlu and M. Zeyveli [18] utilized a genetic algorithm (GA) to optimize the volume of a two-stage helical gearbox. To handle the design constraints, such as bending stress, contact stress, number of teeth on pinion and gear, module, and face width of gear, the objective function was subject to static and dynamic penalty functions. The results from the GA were compared to those of a deterministic design procedure, with the GA being found to be the superior method. D. F. Thompson and colleagues [19] presented a generalized optimization technique for reducing the volume of two-stage and three-stage helical gearboxes, while considering tradeoffs with surface fatigue life. In their study, Edmund S. Maputi and Rajesh Arora [20] explored multi-objective optimization by simultaneously considering three objectives: volume, power output, and center-distance. They employed the NSGA-II evolutionary algorithm to generate Pareto frontiers in their research. From the results of the study, insights for the design of compact gearboxes can be gained. The NSGA-II method was also applied by C. Sanghvi et al. [21] to solve the multi-objective optimization problem of a two-stage helical gearbox for minimum volume and maximum load. The results of the multi-objective optimization of the tooth surface in helical gears using the response surface method were presented by Park C.I. [22].

The best or optimal level of the selected criterion is determined by single-objective optimization, which is an absolute optimization. A multi-objective optimization problem is one with two or more simple goals (or criteria). As a result, the solution to the multi-objective optimization problem cannot best satisfy all criteria simultaneously. For instance, it is not possible to meet both the efficiency and cost requirements of the gearbox. In simple terms, determining a solution to a problem that is both "white" and "black" is impossible; only a "gray" solution can be determined. The gray solution is the one that falls between the best and worst solutions, or between "white" and "black" in the multi-objective optimization problem. As a result, it is known as optimization based on gray relation analysis. The original Taguchi method is used to solve the single-objective problem, and the Taguchi method and gray relation analysis are required to solve the multi-objective optimization problem.

While numerous studies have focused on multi-objective optimization for helical gearboxes, the identification of optimal main design factors for such gearboxes has not received adequate attention. Furthermore, previous research on multi-objective optimization for helical gearboxes has not demonstrated the relationship between optimal input factors and total gearbox ratio. This is a critical issue to consider when designing a new gearbox. In this paper, we present a multi-objective optimization study for a two-stage helical gearbox, considering two single objectives: minimizing gearbox mass and maximizing gearbox efficiency. The proposal of five optimal main design factors for the two-stage helical gearbox is the most significant result of this research. These variables include the CWFW for both stages, the ACS for both stages, and the first stage's gear ratio. Furthermore, by combining the Taguchi method and the GRA in a two-stage process not previously described, we present a novel approach to addressing the multi-objective optimization problem in gearbox design. Additionally, a link between optimal input factors and the total gearbox ratio was proposed.

## 2. Optimization Problem

### 2.1. Gearbox Mass Calculation

The mass of the gearbox $m_{gb}$ can be found by the following equation:

$$m_{gb} = m_g + m_s + m_{gh} + m_b \tag{1}$$

where $m_g$, $m_{gh}$, $m_b$, and $m_s$ designate the mass of the gears, the shafts, the bearings, and the gearbox housing, respectively. The component mass will be specifically calculated below:

### 2.1.1. Gearbox Housing Mass Calculation

The gearbox housing mass ($m_{gh}$) can be found by:

$$m_{gh} = \rho_{gh} \cdot V_{gh} \tag{2}$$

in which $\rho_{gh}$ is the weight density of gearbox housing materials referred; as the material of the gearbox housing is gray cast iron (the most common material for the gearbox housing), $\rho_{gh} = 7300\ (\mathrm{kg/m^3})$; $V_{gh}$ is the gearbox housing volume ($\mathrm{m^3}$), which is determined by the following equation (see Figure 1):

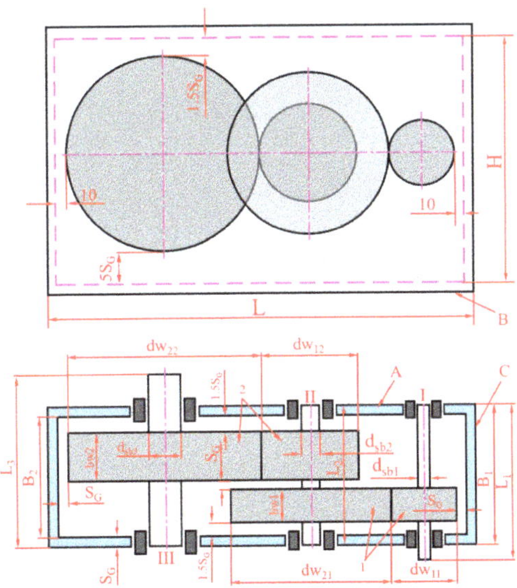

**Figure 1.** Calculated schema.

$$V_{gh} = 2 \cdot V_A + 2 \cdot V_B + 2 \cdot V_C \tag{3}$$

where $V_A$, $V_B$, and $V_C$ are the volumes of side A, B, and C ($\mathrm{m^3}$), respectively.

$$V_A = L \cdot H \cdot S_G \tag{4}$$

$$V_B = L \cdot B_1 \cdot 1.5 \cdot S_G \tag{5}$$

$$V_C = B_2 \cdot H \cdot S_G = (B_1 - 2 \cdot S_G) \cdot H \cdot S_G \tag{6}$$

Substituting Equations (4) to (6) into (3) obtains:

$$V_{gh} = 2 \cdot L \cdot H \cdot S_G + 3 \cdot L \cdot B_1 \cdot S_G + 2 \cdot (B_1 - 2 \cdot S_G) \cdot H \cdot S_G \tag{7}$$

In the above equations, L, H, $B_1$, and SG can be calculated by [2]:

$$L = (d_{w11} + d_{w21}/2 + d_{w12}/2 + d_{w22}/2 + 22.5)/0.975 \tag{8}$$

$$H = d_{w22} + 6.5 \cdot S_G \tag{9}$$

$$B_1 = b_{w1} + b_{w2} + 6 \cdot S_G \tag{10}$$

$$S_G = 0.005 \cdot L + 4.5 \tag{11}$$

2.1.2. Gear Mass Calculation

The gear mass can be determined by:

$$m_g = m_{g1} + m_{g2} \tag{12}$$

in which $m_{g1}$ and $m_{g2}$ are the gear mass of the first and the second stages (kg), which can be calculated by:

$$m_{g1} = \rho_g \cdot \left( \frac{\pi \cdot e_1 . d_{w11}^2 \cdot b_{w1}}{4} + \frac{\pi \cdot e_2 . d_{w21}^2 \cdot b_{w1}}{4} \right) \tag{13}$$

$$m_{g2} = \rho_g \cdot \left( \frac{\pi \cdot e_1 . d_{w12}^2 \cdot b_{w2}}{4} + \frac{\pi \cdot e_2 . d_{w22}^2 \cdot b_{w2}}{4} \right) \tag{14}$$

where $\rho_g$ is the weight density of gear material (kg/m$^3$); as the gear material is steel, $\rho_g = 7800$ (kg/m$^3$); $e_1$ and $e_2$ are the volume coefficients. Because the pinion has a small diameter, its structural form can be plain whereas the gear has a large diameter and thus requires a hub. As a result, the volume coefficient of the pinion $e_1 = 1$, and the volume coefficient of the gear is $e_2 = 0.6$ [13]; $b_{w1}$ and $b_{w2}$ are the gear width of the first and the second stages (mm); $d_{w1i}$ and $d_{w2i}$ are the pinion and the gear pitch diameters of the $i$ stage ($i$ = 1 and 2). These parameters are determined by:

$$b_{w1} = X_{ba1} \cdot a_{w1} \tag{15}$$

$$b_{w2} = X_{ba2} \cdot a_{w2} \tag{16}$$

$$d_{w1i} = 2 \cdot a_{wi}/(u_1 + 1) \tag{17}$$

$$d_{w2i} = 2 \cdot a_{wi} \cdot u_i/(u_i + 1) \tag{18}$$

In the above equations, $u_1$ is the gear ratio of the first stage; the center distance of i stage $a_{wi}$ is found by the surface fatigue strength [23]:

$$a_{wi} = k_a \cdot (u_i + 1) \cdot \sqrt[3]{T_{1i} \cdot k_{H\beta}/(AS_i^2 \cdot u_i \cdot X_{bai})} \tag{19}$$

where $k_a$ is the material coefficient; $k_{H\beta}$ is the contacting load ratio for pitting resistance; $AS_i$ is the allowable contact stress of the i$_{th}$ stage (MPa); $X_{bai}$ is the wheel face width coefficient of ith stage; and $T_{1i}$ is the drive shaft torque of the ith stage (N.mm), which can be calculated by:

$$T_{1i} = \frac{T_r}{\prod_{j=i}^{3}(u_i \cdot \eta_{hg}^{3-i} \cdot \eta_{be}^{4-i})} \tag{20}$$

After calculating the gear parameters, the bending strength of the i$^{\text{th}}$ gear stage must be checked using the following formulas [23]:

$$\sigma_{F1i} = 2 \cdot T_{1i} \cdot K_{Fi} \cdot Y_{\varepsilon i} \cdot Y_{\beta i} \cdot Y_{F1i} / (b_{w1i} \cdot d_{w1i} \cdot m_i) \leq [\sigma_{F1}] \tag{21}$$

$$\sigma_{F2i} = \sigma_{F1i} \cdot Y_{F2i} / Y_{F1i} \leq [\sigma_{F2}] \tag{22}$$

in which $m_i$ is the module of the i$^{\text{th}}$ gear stage (mm); $K_{Fi}$ is load factor; $Y_{\varepsilon i} = 1/\varepsilon$ is the contact factor; $\varepsilon$ is the contact ratio; $Y_{\beta i} = 1 - \beta/140$ is factor taking into account the helix angle; and $Y_{F1i}$ and $Y_{F2i}$ are the geometry factor of the pinion and gear of i$^{\text{th}}$ gear stage.

2.1.3. Calculation of Shaft Mass

The shaft mass can be found by:

$$m_s = \sum_{i=1}^{3} m_{si} \tag{23}$$

where $m_{si}$ is the mass of i$^{\text{th}}$ shaft of the gearbox (kg), which is determined by:

$$m_{si} = \pi \cdot \rho_s \cdot \left( \sum_{j=1}^{n} d_j^2 \cdot l_j + \sum_{k=1}^{2} d_{bk}^2 \cdot B_k \right) \tag{24}$$

in which $d_j$ and $l_j$ are the diameter and the length of j$^{\text{th}}$ shaft part (mm); $d_{bk}$ and $B_k$ are the diameter and the width of k$_{\text{th}}$ shaft part on which the bearing is istalled. The values of $d_j$ and $d_{bk}$ also are determined by [23]:

$$d_j = \sqrt[3]{\frac{M_e}{0.1 \cdot [\sigma_s]}} \tag{25}$$

wherein $[\sigma_s]$ is the allowable shaft stress (MPa), which can be determined by the material and size of the shaft [23]. $M_e$ is the equivalent moment (Nmm), which is found by [23]:

$$M_e = \sqrt{M_y^2 + M_x^2 + 0.75 \cdot T^2} \tag{26}$$

where $M_x$ and $M_y$ are the bending moment in x and y directions (Nmm); T is the torque (Nmm). These parameters can be defined based on the diagram for finding shaft dimensions. Figure 2 describes this diagram for calculation of the first shaft of the gearbox.

In Equation (22), $B_k$ is the bearing width (mm). In this work, radical ball bearings with angular contact were used. From the data in [24], the following regression was proposed to calculate the width of the bearings (with $R^2 = 0.9951$):

$$B_k = 0.0017 \cdot d_{bk}^2 + 0.0277 \cdot d_{bk} + 13.869 \tag{27}$$

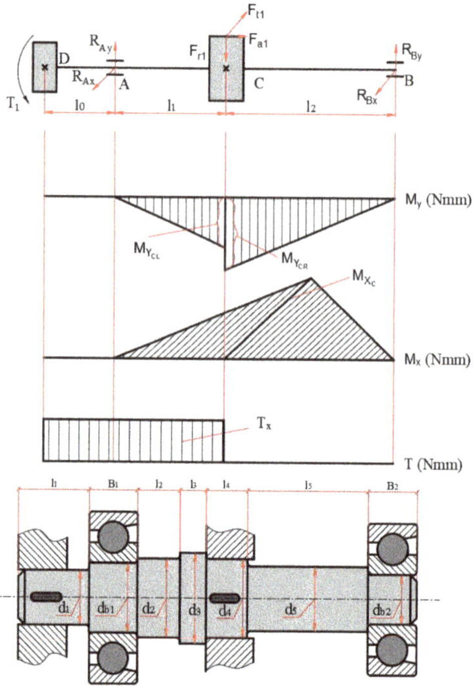



**Figure 2.** Diagram for determining shaft dimensions.

### 2.1.4. Calculation of Bearing Mass

The bearing mass of the gearbox is calculated by:

$$m_b = \sum_{i=1}^{6} m_{bi} \tag{28}$$

As mentioned above, radical ball bearings with angular contact were used in this work. From the data about these type of bearings in [24], the mass of the $i^{th}$ bearing can be found by the following regression equation (with $R^2 = 0.9833$):

$$m_{bi} = 0.0007 \cdot d_{bi}^2 - 0.0513 \cdot d_{bi} + 1.2108 \tag{29}$$

in which i is the number of bearings (i = 6); $d_{bi}$ is the diameter of the shaft part on which the $i^{th}$ bearing is mounted.

### 2.2. Determination of Gearbox Efficiency

The gearbox efficiency is determined by:

$$\eta_{gb} = \frac{100 \cdot P_l}{P_{in}} \tag{30}$$

where $P_l$ is the total power loss in the gear box [25]:

$$P_l = P_{lg} + P_{lb} + P_{ls} \tag{31}$$

where $P_{lg}$ is the power losses in the gears; $P_{lb}$ is the power loss in the bearings; and $P_{ls}$ is the power loss in seals. These factors can be determined as follows:

(+) The power losses in the gears:

$$P_{lg} = \sum_{i=1}^{2} P_{lgi} \tag{32}$$

in which $P_{lgi}$ is the the power losses in the gears of the i stage, which is found by:

$$P_{lgi} = P_{gi} \cdot (1 - \eta_{gi}) \tag{33}$$

where $\eta_{gi}$ is the efficieny of the i stage of the gearbox, which can be determined by [26]:

$$\eta_{gi} = 1 - \left( \frac{1 + 1/u_i}{\beta_{ai} + \beta_{ri}} \right) \cdot \frac{f_i}{2} \cdot \left( \beta_{ai}^2 + \beta_{ri}^2 \right) \tag{34}$$

where $u_i$ is the gear ratio of i stage; $f$ is the friction coefficient; $\beta_{ai}$ and $\beta_{ri}$ are the arc of the approach and recess on the i stage, which is calculated by [26]:

$$\beta_{ai} = \frac{\left( R_{e2i}^2 - R_{02i}^2 \right)^{1/2} - R_{2i} \cdot \sin \alpha}{R_{01i}} \tag{35}$$

$$\beta_{ri} = \frac{\left( R_{e1i}^2 - R_{01i}^2 \right)^{1/2} - R_{1i} \cdot \sin \alpha}{R_{01i}} \tag{36}$$

in which $R_{e1i}$ and $R_{e2i}$ are the outside radius of the pinion and gear, respectively; $R_{1i}$ and $R_{2i}$ are the pitch radius of the pinion and gear, respectively; $R_{01i}$ and $R_{01i}$ are the base-circle radius of the pinion and gear, respectively; $\alpha$ is the pressure angle.

From the data in [26], the friction coefficient can be determined by the folowing regression equations:

-   When the sliding velocity is $v \leq 0.424$ (m/s), the friction coefficient is calculated by (with $R^2 = 0.9958$):

$$f = -0.0877 \cdot v + 0.0525 \tag{37}$$

-   When the sliding velocity is $v > 0.424$ (m/s), the friction coefficient is calculated by (with $R^2 = 0.9796$):

$$f = 0.0028 \cdot v + 0.0104 \tag{38}$$

(+) The power losses in the bearings [25]:

The power losses in rolling bearings can be found by:

$$P_{lb} = \sum_{i=1}^{6} f_b \cdot F_i \cdot v_i \tag{39}$$

where $f_b$ is the coefficient of the friction of the bearing; as the radial ball bearings with angular contact were used, $f_b = 0.0011$ [25]; F denotes the bearing load in Newtons (N) while v represents the peripheral speed. Additionally, i represents the ordinal number of the bearing, ranging from 1 to 6.

(+) The total power losses in the seals are determined by [25]:

$$P_s = \sum_{i=1}^{2} P_{si} \tag{40}$$

in which i is the ordinal number of the seal (i = 1 ÷ 2); $P_{si}$ represents the power loss caused by the sealing for a single seal (w), which can be calculated by:

$$P_{si} = [145 - 1.6 \cdot t_{lil} + 350 \cdot loglog(VG_{40} + 0.8)] \cdot d_s^2 \cdot n \cdot 10^{-7} \tag{41}$$

where $VG_{40}$ is the ISO viscosity grades number.

*2.3. Objective Functions and Constrains*

2.3.1. Objectives Functions

The multi-objective optimization problem in this study comprises two single objectives:

- Minimizing the gearbox mass:

$$\min f_1(X) = \mathrm{m}_{gb} \tag{42}$$

- Maximizing the gearbox efficiency:

$$\min f_2(X) = \eta_{gb} \tag{43}$$

in which X is the design variable vector reflecting variables. In this work, five main design factors, including $u_1$, $Xba_1$, $Xba_2$, $AS_1$, and $AS_2$ were selected as variables, and we have:

$$X = \{u_1, Xba_1, Xba_2, AS_1, AS_2\} \tag{44}$$

2.3.2. Constrains

For a helical gear set, the maximal gear ratio is 9 [23]. Additionally, the coefficient of the wheel face width of both gear stages of a two-stage helical gearbox ranges from 0.25 to 0.4 [23]. In addition, the gear materials used in this work are steel 40, 45, 40X, and 35XM refining, with teeth hardness on the surface of 350 HB (These are the most commonly used gear materials in gearboxes). As a result of the calculated results, the allowable contact stresses of the first and second stages range from 350 to 420 (MPa). Therefore, the following constraints were derived from these comments:

$$1 \leq u_1 \leq 9 \text{ and } 1 \leq u_2 \leq 9 \tag{45}$$

$$0.25 \leq Xba_1 \leq 0.4 \text{ and } 0.25 \leq Xba_2 \leq 0.4 \tag{46}$$

$$350 \leq AS_1 \leq 420 \text{ and } 350 \leq AS_2 \leq 420 \tag{47}$$

**3. Methodology**

As stated in Section 2.3.1 five main design factors were selected as variables for the multi-objective optimization problem. Table 1 describes these factors and the min and max values of them. In this work, the Taguchi method and grey relation analysis were employed to address the multi-objective optimization problem with five variables. In order to easily determine the solutions of the optimization problem, the larger the number of levels of the variables, the better. To maximize the number of levels for each variable, the L25 ($5^5$) design was selected. However, among the variables mentioned, $u_1$ has a very wide distribution (u1 ranges from 1 to 9 as stated in Section 2.3.2). As a result, the gap between levels of this variable remained significant even with five levels (in this case, the gap is $((9-1)/4 = 2)$). To reduce this gap, save time, and improve the accuracy of the solutions, a procedure for solving a multi-objective problem was proposed (Figure 3). This procedure consists of two phases: Phase 1 solves the single-objective optimization problem to close the gap between levels, and phase 2 solves the multi-objective optimization problem to determine the optimal main design factors.

**Table 1.** Input parameters.

| Factor | Notation | Lower Bound | Upper Bound |
|---|---|---|---|
| Gearbox ratio of first stage | $u_1$ | 1 | 9 |
| CWFW of stage 1 | $X_{ba1}$ | 0.25 | 0.4 |
| CWFW of stage 2 | $X_{ba2}$ | 0.25 | 0.4 |
| ACS of stage 1 (MPa) | $AS_1$ | 350 | 420 |
| ACS of stage 2 (MPa) | $AS_2$ | 350 | 420 |

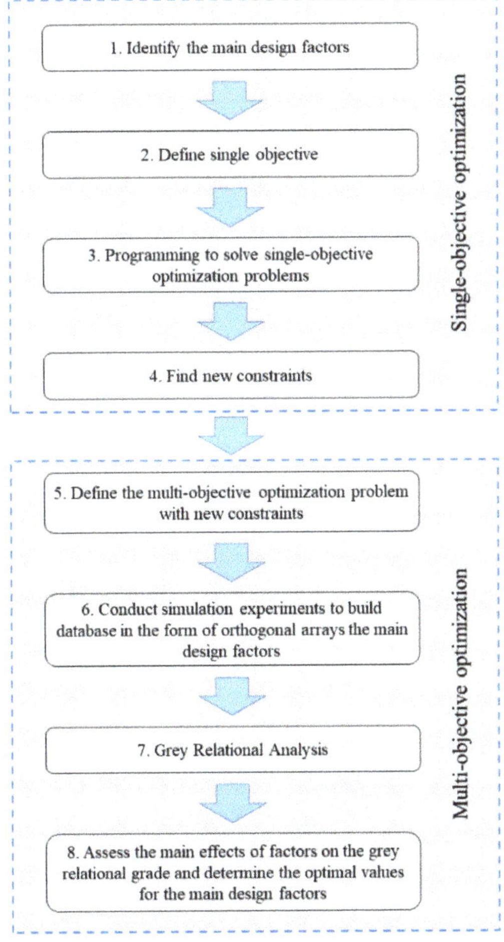


**Figure 3.** The procedure for solving a multi-objective problem.

## 4. Single-Objective Optimization

In this work, the direct search method is used to solve the single-objective optimization problem. Additionally, a computer program has been built using the Matlab language to solve two single-objective problems, including minimizing the gearbox mass and maximizing the gearbox efficiency. From the results of this program, the relation between the optimal value of the gear ratio of the first stage $u_1$ and the total gearbox ratio $u_t$ is shown in Figure 4. Additionally, new constraints for the variable $u_1$ were found, as shown in Table 2.

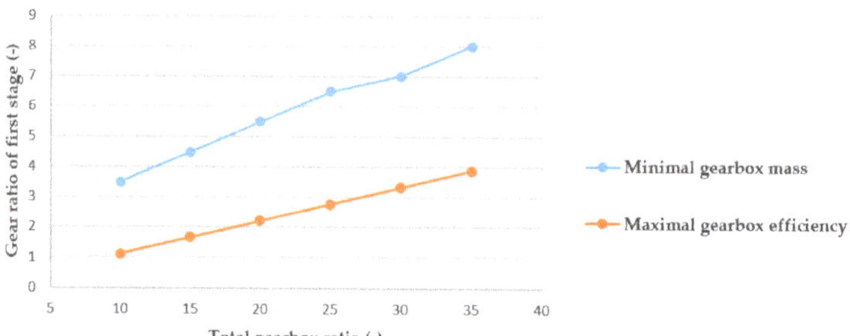


**Figure 4.** Gear ratio of the first stage versus total gearbox ratio.

**Table 2.** New constraints of $u_1$.

| $u_t$ | $u_1$ | |
|---|---|---|
| | Lower Bound | Upper Bound |
| 10 | 1 | 3.6 |
| 15 | 1.5 | 4.6 |
| 20 | 2.1 | 5.6 |
| 25 | 2.7 | 6.6 |
| 30 | 3.2 | 7.1 |
| 35 | 3.8 | 8.1 |

## 5. Multi-Objective Optimization

The multi-objective optimization problem in this research aims to identify the optimal main design factors that satisfy two single-objective functions: minimizing the maximum optimization and maximizing gearbox efficiency in the design of a two-stage helical gearbox with a specific total gearbox ratio. To address this problem, a simulation experiment was conducted. The experiment was designed using the Taguchi method, and the analysis of the results was performed using Minitab R18 software. In addition, as noted above, the design L25 ($5^5$) was chosen for obtaining maximal levels of the variable. A computer program has been developed to perform these experiments. An investigation was conducted to minimize programming intricacy by examining the influence of five key design parameters on gearbox mass. The input first pinion speed of 1480 (rpm) was selected as it is the most common. The steel 45 was selected as the shaft material as it is a very common shaft material. The total gearbox ratios considered for analysis were 10, 15, 20, 25, 30, and 35. Employing a five-level Taguchi design (L25), a total of 25 simulation experiments were carried out for each total gearbox ratio mentioned above. Table 3 describes the main design factors and their levels, and Table 4 presents the experimental plan and the corresponding output results, encompassing the gearbox mass and efficiency, specifically for the total gearbox ratio of 15.

**Table 3.** Main design factors and their levels for $u_t = 15$.

| Factor | Notation | Level | | | | |
|---|---|---|---|---|---|---|
| | | 1 | 2 | 3 | 4 | 5 |
| Gear ratio of first stage | $u_1$ | 1.5 | 2.275 | 3.05 | 3.825 | 4.6 |
| CWFW of stage 1 | $X_{ba1}$ | 0.25 | 0.2875 | 0.325 | 0.3625 | 0.4 |
| CWFW of stage 2 | $X_{ba2}$ | 0.25 | 0.2875 | 0.325 | 0.3625 | 0.4 |
| ACS of stage 1 (MPa) | $AS_1$ | 350 | 367.5 | 385 | 402.5 | 420 |
| ACS of stage 2 (MPa) | $AS_2$ | 350 | 367.5 | 385 | 402.5 | 420 |

**Table 4.** Experimental plans and output responses for $u_t = 15$.

| Exp. No. | Main Design Factors | | | | | $m_{gb}$ (kg) | $\eta_{gb}$ (%) |
| --- | --- | --- | --- | --- | --- | --- | --- |
| | $u_1$ | $X_{ba1}$ | $X_{ba2}$ | $AS_1$ | $AS_2$ | | |
| 1 | 1.500 | 0.2500 | 0.2500 | 350.0 | 350.0 | 106.711 | 98.141 |
| 2 | 1.500 | 0.2875 | 0.2875 | 367.5 | 367.5 | 97.001 | 98.215 |
| 3 | 1.500 | 0.3250 | 0.3250 | 385.0 | 385.0 | 88.818 | 98.192 |
| 4 | 1.500 | 0.3625 | 0.3625 | 402.5 | 402.5 | 81.850 | 98.166 |
| 5 | 1.500 | 0.4000 | 0.4000 | 420.0 | 420.0 | 75.928 | 98.149 |
| 6 | 2.275 | 0.2500 | 0.2875 | 385.0 | 402.5 | 67.773 | 97.882 |
| 7 | 2.275 | 0.2875 | 0.3250 | 402.5 | 420.0 | 62.617 | 97.936 |
| 8 | 2.275 | 0.3250 | 0.3625 | 420.0 | 350.0 | 79.302 | 97.942 |
| 9 | 2.275 | 0.3625 | 0.4000 | 350.0 | 367.5 | 72.893 | 97.918 |
| 10 | 2.275 | 0.4000 | 0.2500 | 367.5 | 385.0 | 74.007 | 97.883 |
| 11 | 3.050 | 0.2500 | 0.3250 | 420.0 | 367.5 | 65.483 | 97.665 |
| 12 | 3.050 | 0.2875 | 0.3625 | 350.0 | 385.0 | 63.794 | 97.781 |
| 13 | 3.050 | 0.3250 | 0.4000 | 367.5 | 402.5 | 59.249 | 97.767 |
| 14 | 3.050 | 0.3625 | 0.2500 | 385.0 | 420.0 | 57.973 | 97.638 |
| 15 | 3.050 | 0.4000 | 0.2875 | 402.5 | 350.0 | 71.479 | 97.658 |
| 16 | 3.825 | 0.2500 | 0.3625 | 367.5 | 420.0 | 54.465 | 97.528 |
| 17 | 3.825 | 0.2875 | 0.4000 | 385.0 | 350.0 | 66.095 | 97.605 |
| 18 | 3.825 | 0.3250 | 0.2500 | 402.5 | 367.5 | 64.069 | 97.461 |
| 19 | 3.825 | 0.3625 | 0.2875 | 420.0 | 385.0 | 58.905 | 97.456 |
| 20 | 3.825 | 0.4000 | 0.3250 | 350.0 | 402.5 | 59.109 | 97.598 |
| 21 | 4.600 | 0.2500 | 0.4000 | 402.5 | 385.0 | 57.246 | 97.369 |
| 22 | 4.600 | 0.2875 | 0.2500 | 420.0 | 402.5 | 55.637 | 97.223 |
| 23 | 4.600 | 0.3250 | 0.2875 | 350.0 | 420.0 | 57.175 | 97.357 |
| 24 | 4.600 | 0.3625 | 0.3250 | 367.5 | 350.0 | 67.270 | 97.296 |
| 25 | 4.600 | 0.4000 | 0.3625 | 385.0 | 367.5 | 62.034 | 97.425 |

The multi-optimization optimization problem is solved by applying the Taguchi and GRA methods. The main steps for this process are as follows:

(+) Determining the signal to noise ratio (S/N) by the following equations as the object of this work is to reduce the gearbox mass and to increase the gearbox efficiency:

- For the gearbox mass objective, the-smaller-is-the-better S/N:

$$SN = -10log_{10}(\frac{1}{n}\sum_{i=1}^{m} y_i^2) \tag{48}$$

- For the gearbox efficiency objective, the-larger-is-the-better S/N:

$$SN = -10log_{10}(\frac{1}{n}\sum_{i=1}^{m} \frac{1}{y_i^2}) \tag{49}$$

where $y_i$ is the ouput response value; m is number of experimental repetitions. In this case, $m - 1$ because the experiment is a simulation; no repetition is required.

The calculated S/N indexes for the two mentioned output targets are presented in Table 5.

**Table 5.** Values of S/N of each experimental run of $u_t = 15$.

| Exp. No. | Main Design Factors | | | | | $m_{gb}$ | | $\eta_{gb}$ | |
| --- | --- | --- | --- | --- | --- | --- | --- | --- | --- |
| | $u_1$ | $X_{ba1}$ | $X_{ba2}$ | $AS_1$ | $AS_2$ | Value (kg) | S/N | Value (%) | S/N |
| 1 | 1.500 | 0.2500 | 0.2500 | 350.0 | 350.0 | 106.711 | −40.5642 | 98.141 | 39.8370 |
| 2 | 1.500 | 0.2875 | 0.2875 | 367.5 | 367.5 | 97.001 | −39.7355 | 98.215 | 39.8436 |
| 3 | 1.500 | 0.3250 | 0.3250 | 385.0 | 385.0 | 88.818 | −38.9700 | 98.192 | 39.8415 |
| 4 | 1.500 | 0.3625 | 0.3625 | 402.5 | 402.5 | 81.850 | −38.2604 | 98.166 | 39.8392 |
| 5 | 1.500 | 0.4000 | 0.4000 | 420.0 | 420.0 | 75.928 | −37.6080 | 98.149 | 39.8377 |
| 6 | 2.275 | 0.2500 | 0.2875 | 385.0 | 402.5 | 67.773 | −36.6211 | 97.882 | 39.8141 |
| 7 | 2.275 | 0.2875 | 0.3250 | 402.5 | 420.0 | 62.617 | −35.9338 | 97.936 | 39.8188 |
| 8 | 2.275 | 0.3250 | 0.3625 | 420.0 | 350.0 | 79.302 | −37.9857 | 97.942 | 39.8194 |
| 9 | 2.275 | 0.3625 | 0.4000 | 350.0 | 367.5 | 72.893 | −37.2537 | 97.918 | 39.8173 |
| 10 | 2.275 | 0.4000 | 0.2500 | 367.5 | 385.0 | 74.007 | −37.3855 | 97.883 | 39.8141 |
| 11 | 3.050 | 0.2500 | 0.3250 | 420.0 | 367.5 | 65.483 | −36.3226 | 97.665 | 39.7948 |
| 12 | 3.050 | 0.2875 | 0.3625 | 350.0 | 385.0 | 63.794 | −36.0956 | 97.781 | 39.8051 |
| 13 | 3.050 | 0.3250 | 0.4000 | 367.5 | 402.5 | 59.249 | −35.4536 | 97.767 | 39.8038 |
| 14 | 3.050 | 0.3625 | 0.2500 | 385.0 | 420.0 | 57.973 | −35.2645 | 97.638 | 39.7924 |
| 15 | 3.050 | 0.4000 | 0.2875 | 402.5 | 350.0 | 71.479 | −37.0836 | 97.658 | 39.7942 |
| 16 | 3.825 | 0.2500 | 0.3625 | 367.5 | 420.0 | 54.465 | −34.7224 | 97.528 | 39.7826 |
| 17 | 3.825 | 0.2875 | 0.4000 | 385.0 | 350.0 | 66.095 | −36.4034 | 97.605 | 39.7894 |
| 18 | 3.825 | 0.3250 | 0.2500 | 402.5 | 367.5 | 64.069 | −36.1330 | 97.461 | 39.7766 |
| 19 | 3.825 | 0.3625 | 0.2875 | 420.0 | 385.0 | 58.905 | −35.4030 | 97.456 | 39.7762 |
| 20 | 3.825 | 0.4000 | 0.3250 | 350.0 | 402.5 | 59.109 | −35.4331 | 97.598 | 39.7888 |
| 21 | 4.600 | 0.2500 | 0.4000 | 402.5 | 385.0 | 57.246 | −35.1549 | 97.369 | 39.7684 |
| 22 | 4.600 | 0.2875 | 0.2500 | 420.0 | 402.5 | 55.637 | −34.9073 | 97.223 | 39.7554 |
| 23 | 4.600 | 0.3250 | 0.2875 | 350.0 | 420.0 | 57.175 | −35.1441 | 97.357 | 39.7673 |
| 24 | 4.600 | 0.3625 | 0.3250 | 367.5 | 350.0 | 67.270 | −36.5564 | 97.296 | 39.7619 |
| 25 | 4.600 | 0.4000 | 0.3625 | 385.0 | 367.5 | 62.034 | −35.8526 | 97.425 | 39.7734 |

In fact, the data of the two considered single-objective functions have different dimensions. To ensure comparability, it is essential to normalize the data, bringing them to a standardized scale. The data normalization is performed using the normalization value Zij, which ranges from 0 to 1. This value is determined using the following formula:

$$Z_i = \frac{SN_i - min(SN_i, = 1, 2, \ldots n)}{max(SN_i, j = 1, 2, \ldots n) - min(SN_i, = 1, 2, \ldots n)} \tag{50}$$

In the formula, "n" represents the experimental number, which in this case is 25.

(+) Calculating the grey relational coefficient:

The grey relational coefficient is calculated by:

$$y_i(k) = \frac{\Delta_{min} + \xi.\Delta_{max}(k)}{\Delta_i(k) + \xi.\Delta_{max}(k)} \tag{51}$$

with $i = 1, 2, \ldots, n$. In the formula, "k" represents the number of objective targets, which is 2 in this case; $\Delta_j(k)$ is the absolute value, $\Delta_j(k) = \|Z_0(k) - Z_j(k)\|$ width; $Z_0(k)$ and $Z_j(k)$ are the reference and the specific comparison sequences, respectively; $\Delta_{min}$ and $\Delta_{max}$ are the min and max values of $\Delta i(k)$; $\zeta$ is the characteristic coefficient, $0 \leq \zeta \leq 1$. In this work $\zeta = 0.5$.

(+) Calculating the mean of the grey relational coefficient:

The degree of grey relation is determined by calculating the mean of the grey relational coefficients associated with the output objectives.

$$\overline{y_i} = \frac{1}{k} \sum_{j=0}^{k} y_{ij}(k) \tag{52}$$

where $y_{ij}$ is the grey relation value of the $j^{th}$ output targets in the $i^{th}$ experiment.

Table 6 displays the calculated results of the grey relation value ($y_i$) and the average grey relation value of all experiments.

**Table 6.** Values of $\Delta_i(k)$ and $\overline{y_i}$.

| Exp. No | S/N | | Zi | | $\Delta_i$ (k) | | Grey Relation Value $y_i$ | | $\overline{y_i}$ |
|---|---|---|---|---|---|---|---|---|---|
| | $m_{gb}$ | $\eta_{gb}$ | $m_{gb}$ | $\eta_{gb}$ | | $\eta_{gb}$ | $m_{gb}$ | $\eta_{gb}$ | |
| | | | Reference Value | | | | | | |
| | | | 1.000 | 1.000 | | | | | |
| 1 | −40.5642 | 39.8370 | 0.0000 | 0.9258 | 1.000 | 0.074 | 0.333 | 0.871 | 0.602 |
| 2 | −39.7355 | 39.8436 | 0.1418 | 1.0000 | 0.858 | 0.000 | 0.368 | 1.000 | 0.684 |
| 3 | −38.9700 | 39.8415 | 0.2729 | 0.9769 | 0.727 | 0.023 | 0.407 | 0.956 | 0.682 |
| 4 | −38.2604 | 39.8392 | 0.3944 | 0.9508 | 0.606 | 0.049 | 0.452 | 0.910 | 0.681 |
| 5 | −37.6080 | 39.8377 | 0.5060 | 0.9338 | 0.494 | 0.066 | 0.503 | 0.883 | 0.693 |
| 6 | −36.6211 | 39.8141 | 0.6750 | 0.6654 | 0.325 | 0.335 | 0.606 | 0.599 | 0.603 |
| 7 | −35.9338 | 39.8188 | 0.7926 | 0.7198 | 0.207 | 0.280 | 0.707 | 0.641 | 0.674 |
| 8 | −37.9857 | 39.8194 | 0.4414 | 0.7258 | 0.559 | 0.274 | 0.472 | 0.646 | 0.559 |
| 9 | −37.2537 | 39.8173 | 0.5667 | 0.7017 | 0.433 | 0.298 | 0.536 | 0.626 | 0.581 |
| 10 | −37.3855 | 39.8141 | 0.5441 | 0.6665 | 0.456 | 0.334 | 0.523 | 0.600 | 0.561 |
| 11 | −36.3226 | 39.7948 | 0.7261 | 0.4468 | 0.274 | 0.553 | 0.646 | 0.475 | 0.560 |
| 12 | −36.0956 | 39.8051 | 0.7649 | 0.5637 | 0.235 | 0.436 | 0.680 | 0.534 | 0.607 |
| 13 | −35.4536 | 39.8038 | 0.8748 | 0.5496 | 0.125 | 0.450 | 0.800 | 0.526 | 0.663 |
| 14 | −35.2645 | 39.7924 | 0.9072 | 0.4196 | 0.093 | 0.580 | 0.843 | 0.463 | 0.653 |
| 15 | −37.0836 | 39.7942 | 0.5958 | 0.4398 | 0.404 | 0.560 | 0.553 | 0.472 | 0.512 |
| 16 | −34.7224 | 39.7826 | 1.0000 | 0.3085 | 0.000 | 0.691 | 1.000 | 0.420 | 0.710 |
| 17 | −36.4034 | 39.7894 | 0.7122 | 0.3863 | 0.288 | 0.614 | 0.635 | 0.449 | 0.542 |
| 18 | −36.1330 | 39.7766 | 0.7585 | 0.2408 | 0.241 | 0.759 | 0.674 | 0.397 | 0.536 |
| 19 | −35.4030 | 39.7762 | 0.8835 | 0.2358 | 0.117 | 0.764 | 0.811 | 0.396 | 0.603 |
| 20 | −35.4331 | 39.7888 | 0.8783 | 0.3792 | 0.122 | 0.621 | 0.804 | 0.446 | 0.625 |
| 21 | −35.1549 | 39.7684 | 0.9260 | 0.1478 | 0.074 | 0.852 | 0.871 | 0.370 | 0.620 |
| 22 | −34.9073 | 39.7554 | 0.9683 | 0.0000 | 0.032 | 1.000 | 0.940 | 0.333 | 0.637 |
| 23 | −35.1441 | 39.7673 | 0.9278 | 0.1357 | 0.072 | 0.864 | 0.874 | 0.366 | 0.620 |
| 24 | −36.5564 | 39.7619 | 0.6860 | 0.0739 | 0.314 | 0.926 | 0.614 | 0.351 | 0.482 |
| 25 | −35.8526 | 39.7734 | 0.8065 | 0.2045 | 0.193 | 0.796 | 0.721 | 0.386 | 0.553 |

To ensure harmony among the output parameters, a higher average grey relation value is desirable. As a result, the objective function of the multi-objective problem can be transformed into a single-objective optimization problem, with the mean grey relation value serving as the output.

The impact of the main design factors on the average grey relation value ($\overline{y}$) was analyzed using the ANOVA method, and the corresponding results are presented in Table 7. From the results in Table 7, $AS_2$ has the most influence on $\overline{y}$ (57.35%), followed by the influence of $u_1$ (25.16%), $X_{ba1}$ (5.38%), $X_{ba2}$ 2.50%), and $AS_1$ (0.84%). The order of influence of the main design factors on $\overline{y}$ through ANOVA analysis is described in Table 8.

**Table 7.** Factor effect on $\bar{y}$.

| Analysis of Variance for Means | | | | | | | |
|---|---|---|---|---|---|---|---|
| Source | DF | Seq SS | Adj SS | Adj MS | F | P | C (%) |
| u1 | 4 | 0.022663 | 0.022663 | 0.005666 | 2.87 | 0.166 | 25.16 |
| Xba1 | 4 | 0.004846 | 0.004846 | 0.001212 | 0.61 | 0.676 | 5.38 |
| Xba2 | 4 | 0.002256 | 0.002256 | 0.000564 | 0.29 | 0.874 | 2.50 |
| AS1 | 4 | 0.000757 | 0.000757 | 0.000189 | 0.10 | 0.978 | 0.84 |
| AS2 | 4 | 0.051656 | 0.051656 | 0.012914 | 6.54 | 0.048 | 57.35 |
| Residual Error | 4 | 0.007900 | 0.007900 | 0.001975 | | | 8.77 |
| Total | 24 | 0.090079 | | | | | |
| Model Summary | | | | | | | |
| | S | | | R-Sq | | | R-Sq(adj) |
| | 0.0444 | | | 91.23% | | | 47.38% |

**Table 8.** Order of main design factor effect on $\bar{y}$.

| Response Table for Means | | | | | |
|---|---|---|---|---|---|
| Level | u1 | Xba1 | Xba2 | AS1 | AS2 |
| 1 | 0.6684 | 0.619 | 0.5978 | 0.6071 | 0.5395 |
| 2 | 0.5956 | 0.6288 | 0.6045 | 0.6202 | 0.5829 |
| 3 | 0.5992 | 0.6119 | 0.6047 | 0.6065 | 0.6148 |
| 4 | 0.6032 | 0.6002 | 0.6222 | 0.6047 | 0.6418 |
| 5 | 0.5827 | 0.5891 | 0.6198 | 0.6105 | 0.67 |
| Delta | 0.0858 | 0.0397 | 0.0243 | 0.0154 | 0.1305 |
| Rank | 2 | 3 | 4 | 5 | 1 |

Average of grey analysis value: 0.610

(+) Determining optimal main design factors:

Theoretically, the set of main design parameters with the levels that have the highest S/N values would be the rational (or optimal) parameter set. Therefore, the impact of the main design factors on the S/N ratio was determined (Figure 5). From Figure 3, the optimal levels and values of main design factors for multi-objective function were found (Table 9).

**Table 9.** Optimal levels and values of main design factors.

| No. | Input Parameters | Code | Optimum Level | Optimum Value |
|---|---|---|---|---|
| 1 | Total gearbox ratio | $u_1$ | 1 | 1.5 |
| 2 | CWFW of stage 1 | Xba1 | 2 | 0.2875 |
| 3 | CWFW of stage 2 | Xba2 | 1 | 0.3625 |
| 4 | ACS of stage 1 (MPa) | AS1 | 5 | 367.5 |
| 5 | ACS of stage 2 (MPa) | AS2 | 5 | 420 |

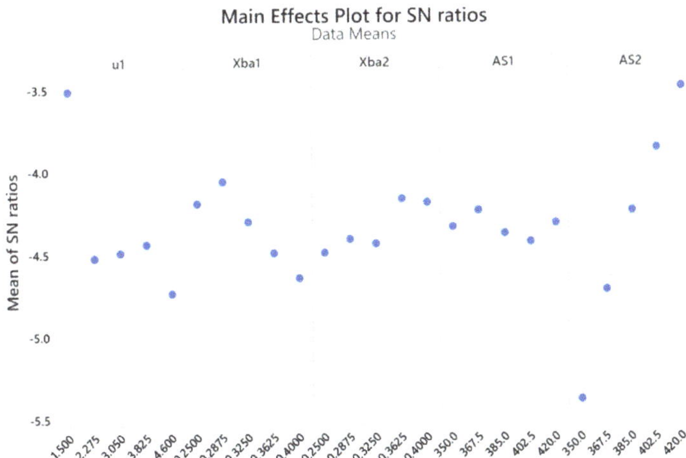



**Figure 5.** Influence of main design factors on S/N ratios of $\overline{y}$.

(+) Evaluation of experimental model:

The adequacy of the proposed model is assessed using the Anderson–Darling method, and the results are presented in Figure 6. From the graph, it is evident that the data points corresponding to the experimental observations (represented by blue dots) fall within the region bounded by upper and lower limits with a 95% standard deviation. Furthermore, the *p*-value of 0.226 significantly exceeds the significance level $\alpha = 0.05$. These findings indicate that the empirical model employed in this study is appropriate and suitable for the analysis.

Continuing from the previous discussion, the optimal values for the main design parameters corresponding to the remaining $u_t$ values of 10, 20, 25, 30, and 35 are presented in Table 10.

Based on the data in Table 10, the following conclusions can be drawn:

The optimal values for $X_{ba1}$ are mostly their minimum values ($X_{ba1} = 0.25$) and for $X_{ba2}$ are the maximum values ($X_{ba2} = 0.4$). This is due to the desire for a low gearbox mass, and because the second gear stage is more loaded than the first, a larger Xba is required to reduce the diameter and thus the total gear mass. This is also consistent with the instructions in [23] for determining the factor $X_{ba}$ for a two-stage helical gearbox.

**Table 10.** Optimum main design factors finding by traditional method.

| No. | $u_t$ | | | | | |
|---|---|---|---|---|---|---|
| | 10 | 15 | 20 | 25 | 30 | 35 |
| $u_1$ | 1 | 1.5 | 2.1 | 2.7 | 3.2 | 3.8 |
| $X_{ba1}$ | 0.25 | 0.2875 | 0.2875 | 0.25 | 0.25 | 0.25 |
| $X_{ba2}$ | 0.4 | 0.3625 | 0.4 | 0.4 | 0.4 | 0.4 |
| $AS_1$ | 350 | 367.5 | 367.5 | 367.5 | 367.5 | 420 |
| $AS_2$ | 420 | 420 | 420 | 420 | 420 | 420 |

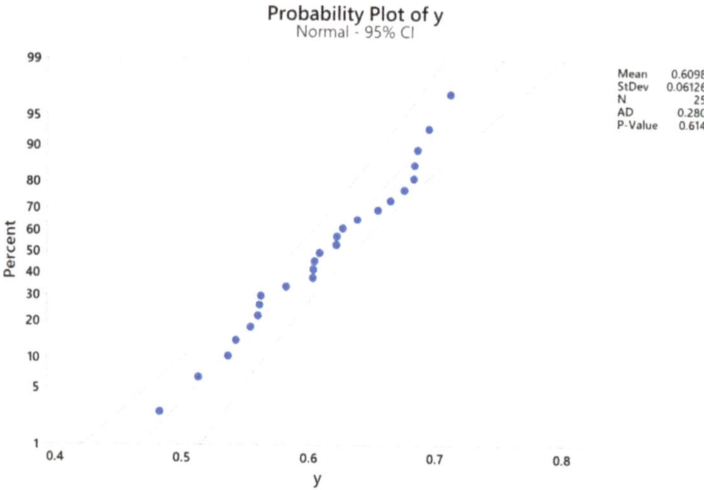

**Figure 6.** Probability plot of $\bar{y}$.

Similarly, the optimal values for $AS_1$ and $AS_2$ are also their maximum values. This is because minimizing the gearbox mass requires maximizing the values of $AS_1$ and $AS_2$. By increasing these values, the center distance of gear stage i (as represented by Equation (19)) can be minimized. It will lead to the gear widths (as determined by Equations (15) and (16)), the pinion, and the gear pitch diameters of the ith stage (i = 1 and 2) (as calculated by Equations (17) and (18)), and therefore, the gear mass (as represented by Equations (13) and (14)) can be minimized.

Figure 7 depicts an obvious first-order relationship among the optimal values of $u_1$ and $u_t$. Additionally, the following regression equation (with $R^2 = 0.9992$) to find the optimal values of $u_1$ was found:

$$u_1 = 0.1126 \cdot u_t - 0.1495 \tag{53}$$

After finding $u_1$, the optimum value of $u_2$ can be determined by $u_2 = u_t / u_1$.

To assess the effectiveness of the proposed method, the multi-objective optimization problem was solved using the constraints of $u_1$, as shown in Table 1 (referred to as the solution by the traditional method). The optimal values for the main design parameters discovered by this method are shown in Table 11. Figure 8 depicts the optimal values of $u_1$ as determined by the traditional method (data from Table 10) and the new method (data from Table 9). This figure clearly shows that the optimal values of $u_1$ for the new method are easily determined and obey a very simple first-order function (Equation (52)). Furthermore, when determined by traditional methods, these values are distributed randomly rather than according to common rules (Figure 8), and they will almost certainly be less accurate than when determined by the new method.

**Table 11.** Optimum main design factors finding by new method.

| No. | $u_t$ | | | | | |
|---|---|---|---|---|---|---|
| | 10 | 15 | 20 | 25 | 30 | 35 |
| $u_1$ | 1 | 1.5 | 2.1 | 2.7 | 3.2 | 3.8 |
| $X_{ba1}$ | 0.25 | 0.2875 | 0.2875 | 0.25 | 0.25 | 0.25 |
| $X_{ba2}$ | 0.25 | 0.25 | 0.25 | 0.25 | 0.25 | 0.25 |
| $AS_1$ | 367.5 | 420 | 420 | 420 | 420 | 420 |
| $AS_2$ | 420 | 420 | 420 | 420 | 420 | 420 |

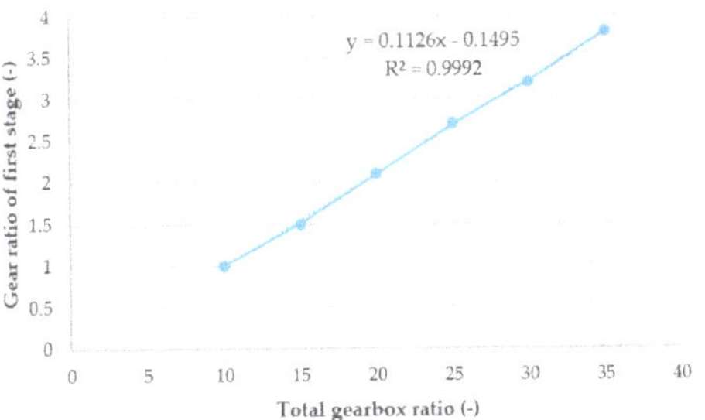



**Figure 7.** Optimal gear ratio of the first stage versus total gearbox ratio.

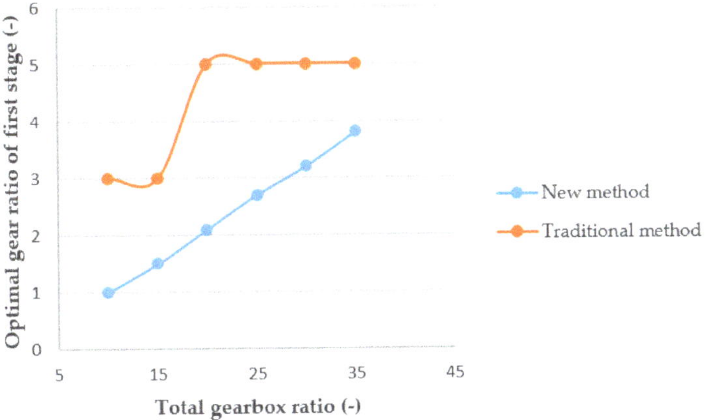



**Figure 8.** Optimal gear ratio of first stage of traditional and new method.

## 6. Conclusions

The Taguchi method and the GRA are used in this paper to solve the multi-objective optimization problem in designing a two-stage helical gearbox. The goal of the study is to discover the optimal main design parameters that maximize gearbox efficiency while minimizing gearbox mass. To accomplish this, five major design factors were chosen: the CWFW for the first and second stages, the ACS for the first and second stages, and the first-stage gear ratio. In addition, the multi-objective optimization problem is solved in two stages. Phase 1 is concerned with solving the single-objective optimization problem of closing the gap between variable levels, while Phase 2 is concerned with determining the optimal main design factors. The following conclusions were proposed as a result of this work:

- A novel approach to handling the multi-objective optimization problem in the gearbox design was presented by combining the Taguchi method and the GRA in a two-stage process. The distance between the values of the lower and upper bounds of the constants of $u_1$ is shortened as a result of this approach, which leads to an easier and a more accurate determination of optimal values.
- The solution of the single-objective optimization problem bridges the gap between variable levels, making the solution of the multi-objective optimization problem easier and more accurate.
- From the results of the study, optimal values for the five main design factors in the design of a two-stage helical gear gearbox were proposed (Equation (52) and Table 10).
- The effect of the main design parameters on $\bar{y}$ was analyzed using the ANOVA method. The results revealed that $AS_2$ had the highest influence on $\bar{y}$ (57.35%), followed by $u_1$ (25.16%), $X_{ba1}$ (5.38%), $X_{ba2}$ (2.50%), and $AS_1$ (0.84%).

- The proposed model of $u_1$ demonstrates a high level of consistency with the experimental data, validating their reliability. This model can be effectively utilized for multi-objective optimization of a two-stage helical gearbox, providing a valuable application.

**Author Contributions:** The original idea was proposed by N.-P.V. The optimization problem was conducted by X.-H.L. and N.-P.V. The manuscript was written by N.-P.V. with support from X.-H.L. Furthermore, both authors were involved in the design of simulations, analysis of experimental figures, and interpretation of experimental results. Finally, N.-P.V. headed the project and revised the paper. All authors have read and agreed to the published version of the manuscript.

**Funding:** This research received no external funding.

**Acknowledgments:** The authors would like to thank Thai Nguyen University of Technology for their assistance with this work.

**Conflicts of Interest:** The authors declare that they have no conflict of interest.

## References

1. Patil, M.; Ramkumar, P.; Shankar, K. Multi-objective optimization of the two-stage helical gearbox with tribological constraints. *Mech. Mach. Theory* **2019**, *138*, 38–57. [CrossRef]
2. Römhild, I.; Linke, H. Gezielte Auslegung Von Zahnradgetrieben mit minimaler Masse auf der Basis neuer Berechnungsverfahren. *Konstruktion* **1992**, *44*, 229–236.
3. Milou, G.; Dobre, G.; Visa, F.; Vitila, H. Optimal design of two step gear units, regarding the main parameters. *VDI Ber.* **1996**, *1230*, 227.
4. Pi, V.N. Optimal calculation of partial transmission ratios for four-step helical gearboxes with first and third step double gear-sets for minimal gearbox length. In Proceedings of the American Conference on Applied Mathematics, Cambridge MA, USA, 24–26 March 2008; pp. 29–32.
5. Pi, V.N. A new study on optimal calculation of partial transmission ratios of two-step helical gearboxes. In Proceedings of the 2nd WSEAS International Conference on Computer Engineering and Applications, CEA, Acapulco, Mexico, 25–27 January 2008; pp. 162–165.
6. Pi, V.N. A new study on the optimal prediction of partial transmission ratios of three-step helical gearboxes with second-step double gear-sets. *WSEAS Trans. Appl. Theor. Mech* **2007**, *2*, 156–163.
7. Danh, T.H.; Linh, N.H.; Danh, B.T.; Tan, T.M.; Van Trang, N.; Thao, T.T.P.; Cuong, N.M. Effect of Main Design Factors on Two-Stage Helical Gearbox Length. In Proceedings of the Advances in Engineering Research and Application: Proceedings of the International Conference on Engineering Research and Applications, ICERA, Thai Nguyen, Vietnam, 1–2 December 2022, pp. 924–932.
8. Pi, V.N. Optimal Calculation of Partial Transmission Ratios of Four-step Helical Gearboxes for Getting Minimal Cross Section Dimension. In Proceedings of the International MultiConference of Engineers and Computer Scientists, Hong Kong, China, 1 January 2008.
9. Pi, V.N. A study on optimal determination of partial transmission ratios of helical gearboxes with second-step double gear-sets. *Int. J. Mech. Mechatron. Eng.* **2008**, *2*, 26–29.
10. Khai, D.Q.; Danh, T.H.; Danh, B.T.; Tan, T.M.; Cuong, N.M.; Hung, L.X.; Pi, V.N.; Thao, T.T.P. Determination of Optimum Main Design Parameters of a Two-Stage Helical Gearbox for Minimum Gearbox Cross-Section Area. In Proceedings of the Advances in Engineering Research and Application: Proceedings of the International Conference on Engineering Research and Applications, ICERA, Thai Nguyen, Vietnam, 1–2 December 2022; pp. 345–353.
11. Pi, V.N. Optimal determination of partial transmission ratios for four-step helical gearboxes with first and third step double gear-sets for minimal mass of gears. In Proceedings of the WSEAS International Conference on Applied Computing Conference, Istanbul Turkey, 27–30 May 2008; pp. 53–57.
12. Danh, T.H.; Huy, T.Q.; Danh, B.T.; Van Trang, N.; Hung, L.X. Optimization of Main Design Parameters for a Two-Stage Helical Gearbox Based on Gearbox Volume Function. In Proceedings of the Advances in Engineering Research and Application: Proceedings of the International Conference on Engineering Research and Applications, ICERA, Thai Nguyen, Vietnam, 1–2 December 2022; pp. 771–779.
13. Vu, N.-P.; Nguyen, D.-N.; Luu, A.-T.; Tran, N.-G.; Tran, T.-H.; Nguyen, V.-C.; Bui, T.-D.; Nguyen, H.-L. The influence of main design parameters on the overall cost of a gearbox. *Appl. Sci.* **2020**, *10*, 2365. [CrossRef]
14. Quang, N.H.; Tu, N.T.; Linh, N.H.; Luan, N.H.; Anh, L.H.; Tuan, N.A.; Giang, T.N.; Pi, V.N. Determining Optimum Gear Ratios for a Four-Stage Helical Gearbox for Getting Minimum Gearbox Cost. In Proceedings of the Advances in Engineering Research and Application: Proceedings of the International Conference on Engineering Research and Applications, ICERA, Thai Nguyen, Vietnam, 1–2 December 2021; pp. 350–364.
15. Danh, T.H.; Huy, T.Q.; Danh, B.T.; Tan, T.M.; Van Trang, N.; Tung, L.A. Determining Partial Gear Ratios of a Two-Stage Helical Gearbox with First Stage Double Gear Sets for Minimizing Total Gearbox Cost. In Proceedings of the Advances in Engineering

Research and Application: Proceedings of the International Conference on Engineering Research and Applications, ICERA, Thai Nguyen, Vietnam, 1–2 December 2022; pp. 376–388.
16. Anh, L.H.; Linh, N.H.; Quang, N.H.; Duc Lam, P.; Anh Tuan, N.; Khac Tuan, N.; Thi Thanh Nga, N.; Ngoc Pi, V. Cost optimization of two-stage helical gearboxes with second stage double gear-sets. *EUREKA Phys. Eng.* **2021**, 89–101. [CrossRef]
17. Wu, Y.-R.; Tran, V.-T. Transmission and load analysis for a crowned helical gear pair with twist-free tooth flanks generated by an external gear honing machine. *Mech. Mach. Theory* **2016**, *98*, 36–47. [CrossRef]
18. Gologlu, C.; Zeyveli, M. A genetic approach to automate preliminary design of gear drives. *Comput. Ind. Eng.* **2009**, *57*, 1043–1051. [CrossRef]
19. Thompson, D.F.; Gupta, S.; Shukla, A. Tradeoff analysis in minimum volume design of multi-stage spur gear reduction units. *Mech. Mach. Theory* **2000**, *35*, 609–627. [CrossRef]
20. Maputi, E.S.; Arora, R. Multi-objective optimization of a 2-stage spur gearbox using NSGA-II and decision-making methods. *J. Braz. Soc. Mech. Sci. Eng.* **2020**, *42*, 477. [CrossRef]
21. Sanghvi, R.C.; Vashi, A.S.; Patolia, H.; Jivani, R.G. Multi-objective optimization of two-stage helical gear train using NSGA-II. *J. Optim.* **2014**, *2014*, 670297. [CrossRef]
22. Park, C.I. Multi-objective optimization of the tooth surface in helical gears using design of experiment and the response surface method. *J. Mech. Sci. Technol.* **2010**, *24*, 823–829. [CrossRef]
23. Chat, T.; Van Uyen, L. *Design and Calculation of Mechanical Transmissions Systems*; Educational Republishing House: Hanoi, Vietnam, 2007; Volume 1.
24. SKF Group, Roller Bearings. 2018. Available online: https://www.skf.com/binaries/pub12/Images/0901d196802809de-Rolling-bearings{-}{-}-17000_1-EN_tcm_12-121486.pdf (accessed on 24 May 2023).
25. Jelaska, D.T. *Gears and Gear Drives*; John Wiley & Sons: Hoboken, NJ, USA, 2012.
26. Buckingham, E. *Analytical Mechanics of Gears*; Dover Publications, Inc.: New York, NY, USA, 1988.

*Article*

# Vibration-Based Detection of Bearing Damages in a Planetary Gearbox Using Convolutional Neural Networks

**Julia Scholtyssek *, Luka Josephine Bislich, Felix Cordes and Karl-Ludwig Krieger**

Institute of Electrodynamics and Microelectronics, University of Bremen, Otto-Hahn-Allee 1, 28359 Bremen, Germany
* Correspondence: julia.scholtyssek@uni-bremen.de

**Abstract:** Tapered roller bearings are used partly in very rough and highly stressful environmental conditions. Therefore, the need for condition monitoring is increasing. This study is intended to provide an approach for monitoring bearings in a two-stage planetary gearbox based on vibration analysis. In total, the data of six damage phenomena and one healthy bearing are collected. A convolutional neural network (CNN) is trained and evaluated by using the balanced accuracy. Mainly, it is investigated how many damage severities can be detected. In addition, the robustness of the model regarding unknown speeds and damage phenomena should be proven. The results show a very good differentiation up to all of the presented damage phenomena. The classifier reaches an averaged balanced accuracy of 0.96. Also, samples collected at unknown speeds can be classified well for speed values within the known range. For unknown damage phenomena, the classifier shows limits so that a reliable classification is only applicable with a binary classifier, which differentiates between healthy and damaged. The investigations therefore show that a reliable detection of bearing damage is possible in a two-stage planetary gear. Furthermore, the transferability of the model is successfully tested and implemented for the binary classifier.

**Keywords:** planetary gear; bearing damage; vibration analysis; CNN; classification

**Citation:** Scholtyssek, J.; Bislich, L.; Cordes, F.; Krieger, K.-L. Vibration-Based Detection of Bearing Damages in a Planetary Gearbox Using Convolutional Neural Networks. *Appl. Sci.* **2023**, *13*, 8239. https://doi.org/10.3390/app13148239

Academic Editors: Aleksandar Miltenovic, Franco Concli and Stefan Schumann

Received: 26 June 2023
Revised: 12 July 2023
Accepted: 15 July 2023
Published: 16 July 2023

## 1. Introduction

In the present application, the tapered roller bearings are installed in a two-stage planetary gearbox in port vehicles. Since the bearings are exposed to a highly stressful environment, the bearings can evolve the wearing phenomena. Particularly in the port application, unpredictable machine failures must be avoided, as this results in high costs. Therefore, the condition monitoring of the bearings is necessary to detect damage at an early stage. To implement this, acoustic signals are evaluated in this contribution.

There is already a large number of publications regarding the condition monitoring of bearings. The most important results are introduced briefly. In general, two diverging approaches can be distinguished, namely signal processing methods and intelligent methods [1,2]. In general, the signals in the time domain, frequency domain or time–frequency domain can be used for signal-processing methods. Typical methods to analyze the signals are statistical indicators, the detection of characteristic fault frequencies for each component or occurring sidebands and spectrum observations [3,4]. In addition to common methods like Fourier transformation, multiple decomposition methods like empirical mode decomposition or variational mode decomposition, have been developed [5]. The advantage of these methods compared to Fourier transformation is that they are applicable to non-stationary signals, which are frequently occurring in gearboxes. Therefore, the vibration signals are decomposed into mono-component modes. A new method for detecting bearing damage in this field is feature mode decomposition (FMD) [6]. Based on the use of a finite-impulse response filter, the signals are preprocessed. Afterwards, a mode selection based on a combined argument of correlated kurtosis and the correlation coefficient is performed. FMD can successfully detect individual damages and, in contrast to previous decomposition methods, is also capable of extracting the modes in the case of combined damages. The database of all presented signal processing methods can be used as an input for a variant of intelligent methods, like support vector machines or neural networks [7].

Based on the literature, the presented methods are not applicable to every use case. Most of the publications concern laboratory investigations on bearings. Investigations of bearings integrated in a gearbox are rare. Planetary gearboxes are used in rough and variable conditions. Therefore, the evaluation of fault frequencies based on the fast Fourier transformation is not possible since it is designed for stationary signals [8]. Furthermore, planetary gearboxes, and, therefore, bearings in planetary gearboxes, show additional challenges like the superposition of vibration events, nonlinear transmission paths, low-speed components and noise. Also, most investigations are based on localized faults, like local pittings. Investigations on distributed wearing phenomena are rare.

This paper presents an investigation of localized and distributed defects. The investigations on the tapered roller bearings are being developed in the compound system of the two-stage planetary gearbox. In addition, this paper investigates the influence of unknown damage and is not limited to the detection of damage classes which the network has already used in the training process. For this, a convolutional neural network is developed, which takes spectral data as input. The first investigation addresses the class granularity. It is analyzed how many classes the classifier can reliably differentiate. Therefore, a binary classifier differentiating between two classes, healthy and damaged, is first developed. Afterwards, a classifier differentiating between three classes, healthy, localized and distributed damage, is used. Finally, a classifier differentiating between seven classes, which show different severities of the presented subclasses, is designed and evaluated. The second investigation addresses the impact of different speed rotations. The last investigation deals with unknown damage phenomena. For this, the classifier only differentiating in three classes is used. Single damage phenomena are held back during training and are only used in the test set to reveal if a correct classification is possible.

The article is structured as follows. Section 2 presents the experimental setup. First, the two-stage planetary gearbox design is introduced. Afterwards, the investigated damages and lastly, the data acquisition and test rig are presented. Section 3 deals with the dataset and classifier preparation. Afterwards, the different investigations are presented in Section 4. In the first investigation the implementable class granularity is proven. Secondly, the influence of different speed rotations is observed. Finally, the robustness of the developed classifier is tested by classifying unknown damage events. In Section 5, the results are discussed and a conclusion is drawn including the optimization potential.

## 2. Materials and Methods

This section describes the fundamental setup of this contribution, which is used to realize the condition monitoring of bearing damages by using vibration analysis. First, the structure of the examined two-stage planetary gearbox is presented. Afterwards, the occurrence of bearing damages and the evaluated damages are described. Lastly, the experimental setup is outlined.

### 2.1. Two-Stage Planetary Gearbox

The most important specifications of the used two-stage planetary gear should be presented. The gearbox has a maximum input speed rotation of 4170 rpm and a maximum input torque of 450 Nm. The specifications of each stage are presented in Table 1.

**Table 1.** Specifications of the investigated two-stage planetary gearbox.

| Property | First Stage | Second Stage |
|---|---|---|
| number of teeth sun gear | 15 | 12 |
| number of teeth planet gear | 31 | 23 |
| number of teeth ring gear | −78 | −60 |
| number of planets | 3 | 4 |

The total gear ratio is $i = 37.2$. A planetary gear consists of three main components, namely the ring gear, sun gear and planet gears. In total, there are three types of planetary gearbox designs [9]. They differ in the component that is fixed and therefore not rotating. The presented design is based on a fixed ring gear. The sun gear is in the center, rotating around its own center. The planets are placed between the sun and the fixed ring gear and mesh with both components [10,11]. They rotate orbitally to the sun and, in addition, they rotate around their own center. The used planetary gearbox is a two-stage design. Hence, the planet gears of the first stage are connected to the sun of the second stage by using a planet carrier shaft. The construction design is shown in Figure 1.

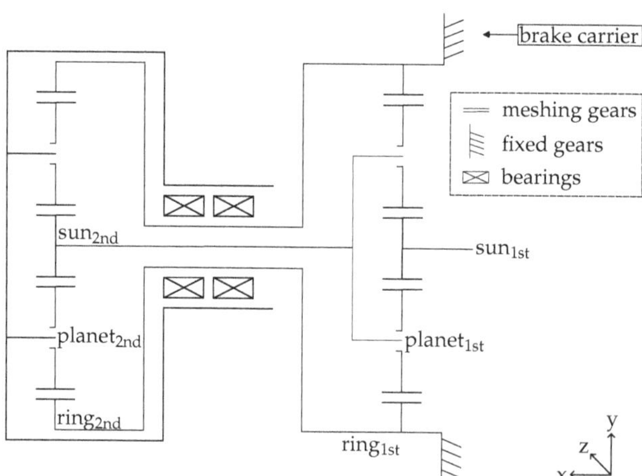


**Figure 1.** Construction draft of a two-stage planetary gear box.

In addition to the gearing components, the two-stage planetary gearbox has two sets of taper roller bearings. Both are mounted between the ring gear of the second stage and the ring gear carrier as depicted in Figure 1. The most important specifications of the applied taper roller bearings are listed in Table 2.

**Table 2.** Specifications of the investigated taper roller bearings.

| Dimensions | | Performance Data | |
|---|---|---|---|
| Inner diameter | 150 mm | Basic dynamic load rating, radial $C_r$ | 455,000 N |
| Diameter between two opposite rolling elements | 187.8 mm | Basic static load rating, radial $C_{0r}$ | 680,000 N |
| Outer diameter | 225 mm | Limiting speed $n_G$ | 3550 1/min |
| Width | 48 mm | Operating temperature range | −30–200 °C |

A taper roller bearing consists of an outer ring, taper rollers and an inner ring. In the presented application, the outer ring is rotating and the inner ring is stationary. Since both bearing sets are rotating with dependence on the wheel hub, they are investigated in conjunction. For each bearing component, an individual frequency can be calculated. The first factor is a characteristic constant for each component, which shows design dependencies. The second factor is the rotation frequency of the wheel hub $f_n$, which depends on the input rotation speed:

$$\text{Ball spin frequency } BSF = BSFF \cdot f_n = 5.06 \cdot f_n, \tag{1}$$

$$\text{Ball pass-frequency outer race } BPFO = BPFFO \cdot f_n = 13.58 \cdot f_n, \tag{2}$$

$$\text{Ball pass-frequency inner race } BPFI = BPFFI \cdot f_n = 16.42 \cdot f_n, \tag{3}$$

$$\text{Fundamental train frequency } FTF = FTFF \cdot f_n = 1.09 \cdot f_n. \tag{4}$$

Planetary gearboxes are gaining increasing attention. They are used in transportation systems, aerospace, automotive applications and wind turbines [9,12]. The main advantages are a compact structure, a stable transmission, a large transmission ratio and high efficiency. Due to rough conditions, the bearings and gearings often show faults like wear, fatigue cracks or pittings. The resulting faulty gearbox can cause operation hazards, with a shutdown as the worst-case scenario and the associated high costs. But some difficulties arise for vibration monitoring. For a two-stage planetary gearbox, the number of meshing components as well as that of the transmission paths increase. As a result, the effects of dispersion, absorption and superposition become more intense. This contribution focuses on the detection of bearing damages in a two-stage planetary gearbox.

*2.2. Bearing Damage*

Generally, occurring bearing damages can be categorized into localized and distributed damages [13,14]. Localized faults can result from material defects. Those faults can be spread over time through particles in the oil circulation system or contact between components like meshing gears. Therefore, it is necessary to detect those faults in an early stage to avoid a progressive fault and a premature failure of the whole system. Distributed faults mostly occur beyond the guaranteed lifetime. They are caused by material fatigue due to the applied load and the tough operating conditions.

This contribution considers three different kinds of damages for each subclass. A comprehensive overview is shown in Table 3. The three bearing damages which belong to the distributed subclass show severe pittings distributed over the outer race and tapper rollers. They only show slight differences in severity. All three damages occurred during real operation.

The three bearing damages, which belong to the localized subclass, are sequentially more severe. They are added with one local fault in each iteration. Therefore, the damage of the class *local 0* is included in classes *local 1* and *local 2*. All three damage phenomena are synthetically produced and are most likely to be compared to scoring. The first damage is severe local damage on the outer race. The second damage is light local damage on the outer race with 9 cm distance to the severe local damage. The last damage is severe damage on one tapper roller. Both the distributed and the localized classes do not have information about the inner race condition. This is due to the fact that an evaluation of the inner race would need a total destruction of the whole bearing set.

**Table 3.** Overview of the considered damaged bearings.

| Subclass | Class | Damaged Components |
|---|---|---|
| healthy | healthy | |
| distributed | dist 0 | severe pittings on outer race and tapper rollers |
| | dist 1 | severe pittings on outer race and tapper rollers |
| | dist 2 | severe pittings on outer race and tapper rollers |
| localized | local 0 | severe scoring on outer race |
| | local 1 | severe and light scoring on outer race |
| | local 2 | severe and light scoring on outer race, severe scoring on tapper roller |

### 2.3. Experimental Setup

For the acquisition of the vibration signals, a piezoelectric vibration transducer is used, which has a linear frequency range from 0.13 Hz to 22 kHz and a sensitivity of 100 mV/g. The sensor is an accelerometer. A signal acquisition at the bearings is not possible. Therefore, the sensor signal acquisition is carried out at the brake carrier of the wheel drive. The brake carrier is located at the first stage as shown in Figure 1. The sensor is mono-axial and orientated in x-direction. The IEPE signal of the sensor is logged through a data-acquisition system with a sampling rate of 100 kHz and a resolution of 24 bit.

The data acquisition is carried out with a load-free test rig. In total, seven measurement campaigns are executed. In the first campaign, an undamaged test gearbox is used. Afterwards, the damaged components listed in Table 3 are separately mounted to the test gearbox. For each campaign, multiple constant rotational speeds $n$ are applied, namely 700, 1200, 1700, 2100, 2700, 3400, 4200 rpm. Each rotational speed is measured five times with a duration of 20 s, respectively.

### 3. Data Processing

In the present investigation, it is very difficult to derive precise predictions of the bearing condition based on signal-processing methods due to the noisy environment. Therefore, intelligent methods are used. A convolutional neural network (CNN) is tested for this purpose. Since the CNN processes image data, spectrograms are generated from the raw signals in the first step. This is explained in more detail in Section 3.1. The general functionality and the structure of the used CNN is presented in Section 3.2. Furthermore, since many classifiers show the overfitting and drawbacks of imbalanced datasets, suitable evaluation methods are identified in Section 3.3 to enable a reliable evaluation of the results.

### 3.1. Data Preparation

Each dataset is divided into data segments with the length of half a second, 50,000 samples respectively, which are transformed into the time–frequency domain and depicted as spectrograms. The spectrograms are scaled logarithmic and the color scale is normed with the minimal and maximal values over the whole dataset. The spectrograms are saved as pictures without an axis because they are normed. The pictures are then used as three-dimensional input for a CNN. In total, the dataset consists of 9800 spectrograms, which are evenly distributed over all seven classes. An example of the generated spectrograms for a rotational speed of 2700 rpm is shown in Figure 2. For the same operating conditions, the resulting spectrograms of three damage cases, respectively *healthy*, *dist 2* and *local 0*, are shown. Slight differences can be detected. In relation to the healthy condition, the signal amplitude in the lower frequency range increases in the case of distributed damage, while the middle frequency range obtains higher amplitudes only within a clearly definable frequency band. For local damage, on the other hand, the lower frequency band has higher amplitude than for the healthy state and is larger. In addition, this damage exhibits higher amplitude–frequency bands in the middle-frequency range. In the high-frequency range, all three damage phenomena behave similarly.

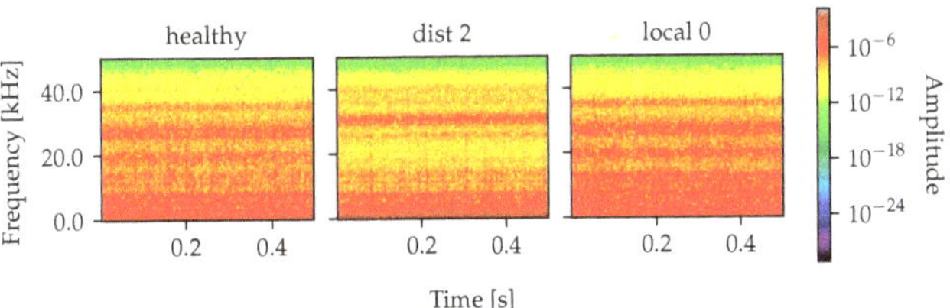

**Figure 2.** Spectrograms of three damage classes for 2700 rpm, which are used as input for the convolutional neural network (CNN).

### 3.2. Classifier

CNNs are used in many applications, which can be summarized mainly into three categories [15]. The first category is classification. Examples for classification are mainly medical image analysis and real-time target tracking. CNNs are also used for object detection, which is the second category. Object detection includes object localization and object classification. It is often used in medical context, for example, in radiology with X-ray images to detect fractured bones or in the case of tumor detection [16]. Lastly, the third main application is segmentation. In addition, CNNs can be used as a preprocessing step for feature extraction [17].

CNNs provide many advantages. They offer to extract features from the raw data with no need for preprocessing steps [17]. Furthermore, they are precise and result in high accuracy [18]. They have a high ability to work with variable or complex data, outperform other intelligent methods, are computationally efficient and only need few parameters [19]. Therefore, the application to noisy acoustic signals in the port environment is promising for the realization of a reliable classifier.

Generally, a CNN is a multi-layer neural network and used for image data [17]. CNNs use discrete convolution to extract information from the image data. The general structure of CNNs is presented shortly. CNNs basically consist of an input layer, hidden layers and an output layer. The hidden layers are convolutional, pooling or fully connected layers. First, the images are fed to the network using the input layer. Afterwards, a combination of multiple convolutional and pooling layers is used to extract features. One convolutional layer is built by a set of multiple kernels [15]. Each kernel has weights, which are window-wise convolved with the input array. This creates features and also reduces the data size [20]. The key parameters are the size and the number of kernels [16], whereas pooling layers are only used to reduce the feature size, while retaining the relevant information. They reduce the dimensionalities and downscale the matrix. Pooling layers, in addition, improve the net transferability to unknown data [18]. After multiple convolutional and pooling layers, the two-dimensional feature space is flattened to one dimensionality. This is performed through a fully connected layer and is required for the classification. At least one layer is necessary but multiple are possible. The last layer has the same size as the defined classes. In each fully connected layer, each input is connected and weighted to the next output. Lastly, a function of activation is needed, which normalizes the output to class probabilities.

CNNs have successfully been used in bearing condition monitoring [2]. Guo et al., for example, used an adaptive deep CNN to differ between four bearing health states [19]. Qian et al. used adaptive overlapping convolutional neural networks to increase the performance by using the one-dimensional raw vibration signal [21]. This method reached high accuracy for a public bearing dataset with ten health conditions, while using only 5 % of the data for training. In addition, the results were verified with a second dataset of a different bearing investigation.

For the presented application, a CNN is developed with tensorflow and tested for differentiating into the presented bearing damage cases. The architecture is shown in Figure 3. In addition to the layer configuration, the two-dimensional kernel size of each layer is provided. Also, the number of features is shown below each convolutional layer.

The first convolutional layer restructures the three-dimensional input, which is comprised of the RGB values of the spectrograms. The layer is combined with a relu activation layer and a max pooling layer. Afterwards, three convolutional layers, each combined with a relu activation layer, a normalization layer and a pooling layer using max pooling, are implemented. Finally, the data are flattened by global pooling, which is connected to a dense layer. Lastly, a second dense layer combined with a softmax activation layer is connected, which has the same size as the considered number of classes. Since the number of classes varies in different investigations, the last layer is variable.

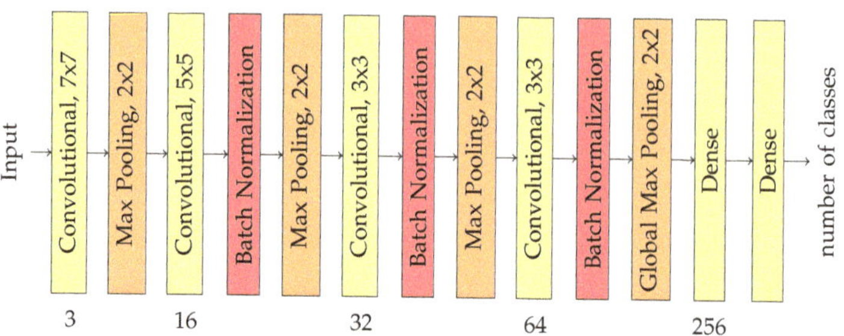


**Figure 3.** Applied CNN architecture.

*3.3. Evaluation*

To train the CNN, k-fold cross validation is used. Typically, in machine and deep learning, $k = 10$ is chosen as in the present investigation. First, the dataset is randomly shuffled. Afterwards, the dataset is divided into ten equal-sized parts, where each part is used once for testing [22,23]. The remaining data are used to develop the CNN-model, where eight parts are used for training and one part is used for validation. As consequence, the model is trained ten times. To evaluate the model accuracy, the mean result as well as the best and the worst iterations are considered. Cross validation ensures that all data points can be used for training and testing without resulting in overfitting [24]. Without a good evaluation, many models tend to classify the trained data perfectly but fail for new data. This is called the generalization problem, which is overcome by using cross validation.

The present investigation is a multi-class problem. The most common classifier metric is the accuracy (ACC). It represents the ratio of correctly classified samples to the whole dataset [25]:

$$ACC = \frac{TP + TN}{TP + FP + TN + FN}.$$

(5)

Therefore, the correctly classified samples of the positive class (TP) and the correctly classified samples of the negative class (TN) are set in relation to all sample points. This includes the correct classification as well as the false classification of the positive (FP) and negative (FN) class.

This metric is useful if there is either a balanced class distribution or only the probability of a correct classification is important, independent of the classes. In the present case, a balanced class distribution is only given when differentiating in seven classes. But this might not be the case for each cross validation. In addition, further information about the ability to detect every single class is needed. Therefore, another metric, handling imbalanced datasets, should be used.

A common solution for this is the balanced accuracy (BAC) [26,27]. The BAC is the average over each class dependent recall:

$$BAC = \frac{1}{N} \sum_{i=1}^{N} recall_i.$$

(6)

The recall is the ratio of correctly classified samples of a defined class. Defining the considered class as positive and all others as negative, the class specific recall is the TP classified samples of a class in relation to all suggested class samples, which includes the TP but also the FN:

$$\text{recall}_i = \frac{\text{TP}}{\text{TP} + \text{FN}}. \tag{7}$$

## 4. Results

The main aim of this contribution is to realize a reliable classifier to detect bearing damages. In this context, some specific investigations are presented for this application. First of all, the implementable class granularity should be tested. This is presented in Section 4.1. The investigations on the test rig are examined on defined speeds. In real operations, constant and defined speeds at any time cannot be assumed. Therefore, the effects of data collected at unknown speeds on the classifier results should be investigated. This is shown in Section 4.2. Lastly, based on the presented class granularities, it is investigated if unknown damages can be sorted into subclasses by the trained classifier. This is shown in Section 4.3. These last two tests are intended to provide an indication of the robustness of the designed classifier from Section 4.1.

### 4.1. Class Granularity

The first study examines whether the detection of bearing damage in the present application is implementable with the use of a CNN. At first, a binary classifier is developed, which distinguishes between healthy and damaged. The second class is composed of the distributed and localized classes. Then, the classification problem is extended to three classes. Thereby, the subclasses healthy, distributed and localized are differentiated. As shown in Table 3, the classes *dist 0*, *dist 1* and *dist 2* are combined for the distributed subclass. The subclass localized consists of *local 0*, *local 1* and *local 2*. In the last step, the class granularity is increased to seven classes. Therefore, all classes shown in Table 3 should be detected. The investigation of the class granularity should provide information as to which granularity should be preferred in the case of a suitable trade-off between high-quality and high-information content. The results are shown in Table 4. Each model is trained ten times using tenfold cross validation, with the averaged BAC over all iterations and the results of the best and worst iterations listed, respectively.

**Table 4.** Classifier results for different class granularities.

| Class Granularity | $\text{BAC}_{\text{mean}}$ | $\text{BAC}_{\text{max}}$ | $\text{BAC}_{\text{min}}$ |
|:---:|:---:|:---:|:---:|
| 2 | 0.99 | 1 | 0.98 |
| 3 | 0.98 | 0.99 | 0.97 |
| 7 | 0.96 | 1 | 0.70 |

The result shows that the binary classifier achieves the overall best result with an average BAC of 0.99. The worse iteration is achieving a BAC of about 0.98. But the classifiers with higher granularity show very good results, too. With increasing class granularity, the BAC decreases. But the classifier with seven classes still reaches an average BAC of 0.96. The worst iteration produces a BAC of 0.70, which is far worse than the other granularities. However, this iteration could be seen as an outlier since the second-worst iteration reaches a BAC of about 0.98. In Figure 4, the result of the worst iteration of the classifier with seven classes is shown. A confusion matrix (CM) is used for visualization. On the horizontal axis, the predicted labels are shown, whereas on the vertical axis, the true labels are depicted. A good classifier should predict the true labels, and therefore most of the samples should be placed on the main diagonal. The CM shows that the distributed classes are classified nearly perfectly, while the healthy and all of the local classes show multiple misclassifications. It is noteworthy that all four classes are often predicted as class *dist 0*.

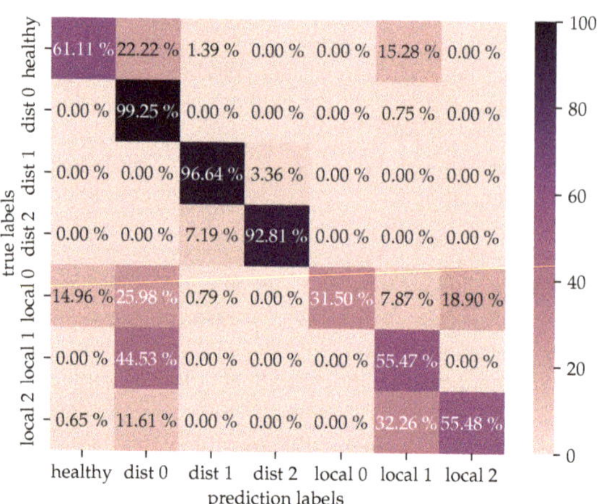


**Figure 4.** Confusion matrix for the worst iteration of the classifier differentiating seven classes.

In summary, the classification of all presented granularities is possible with sufficiently good accuracy. Depending on the requirements and the use case, it should be decided individually which misclassifications are acceptable, and the largest granularity that meets these criteria should be selected accordingly. For the presented use case, the result of the classifier with seven classes is sufficient.

*4.2. Speed Dependency*

The second study examines the influence of speed dependencies. As presented, the data are collected at a constant speed with seven different rotational steps. In the real application, a constant speed at defined values is not possible. Therefore, the influence of the speed on the classifier accuracy should be investigated.

Two investigations are performed. In the first investigation, the classifier is trained with rotational speeds of 700, 1700, 2700 and 4200 rpm and tested with the other three speeds. In the second investigation, the classifier is trained with rotational speeds of 1200, 2100 and 3400 rpm. The following equations provide an overview of the investigations, where $D$ is the whole dataset:

$$\text{Investigation 1: } D_{\text{train}} = D \setminus D_{\text{test}}, \quad D_{\text{test}} = \{1200, 2100, 3400\} \text{ rpm,} \tag{8}$$

$$\text{Investigation 2: } D_{\text{train}} = D \setminus D_{\text{test}}, \quad D_{\text{test}} = \{700, 1700, 2700, 4200\} \text{ rpm.} \tag{9}$$

Again, all investigations are performed using tenfold cross validation. The results are shown in Table 5. The highlighted data show the results of the tested data, which means the trained classifier does not know those rotational speeds. First of all, the BAC is decreasing for samples, if the present rotational speed is not used in the training process. In the worst case, data samples collected at 700 rpm only reach an averaged BAC of 0.20, and the worst iteration only reaches 0.06. Therefore, the classifier becomes worse than random prediction. Secondly, it can be derived from the results that, compared among rotational speeds not used in the training process, the intermediate rotational speeds can achieve the best classification accuracy. Therefore, the data collected at 2100 rpm still reach an averaged BAC of 0.83 in the first investigation, although this speed is not trained. The second and third best results are reached at 3400 and 1700 rpm. Also, it can be noted that the results from the first investigation achieve higher BAC. The main difference to the second investigation is that only intermediate speeds are tested and wrapped by trained rotational speeds. In addition, with the worst results of the lowest and highest rotational speeds and the best results for the middle rotational speeds, it can be derived that for a good classification of unknown rotational speeds, a suitable coverage of rotational speeds during training is essential. Therefore, the rotational speeds in the training set should cover the whole range and be evenly distributed since the investigation shows that data of rotational speeds, which are wrapped by trained rotational speeds, result in good accuracy.

**Table 5.** Classifier results for withheld speeds.

| $n$ | Investigation 1 | | | Investigation 2 | | |
|---|---|---|---|---|---|---|
| | $\mathrm{BAC_{mean}}$ | $\mathrm{BAC_{max}}$ | $\mathrm{BAC_{min}}$ | $\mathrm{BAC_{mean}}$ | $\mathrm{BAC_{max}}$ | $\mathrm{BAC_{min}}$ |
| 700 | 0.99 | 1 | 0.99 | **0.20** | **0.33** | **0.06** |
| 1200 | **0.63** | **0.76** | **0.44** | 0.63 | 0.76 | 0.44 |
| 1700 | 1 | 1 | 0.99 | **0.71** | **0.92** | **0.61** |
| 2100 | **0.83** | **0.90** | **0.73** | 0.99 | 1 | 0.91 |
| 2700 | 1 | 1 | 0.99 | **0.67** | **0.78** | **0.50** |
| 3400 | **0.75** | **0.92** | **0.53** | 0.99 | 1 | 0.95 |
| 4200 | 1 | 1 | 1 | **0.38** | **0.52** | **0.28** |

It can, therefore, be assumed that the ranges in between the known rotational speeds from the present measurements are also classifiable.

### 4.3. Unknown damages

In the third study, the robustness of the classifier is evaluated with regards to unknown damage conditions. To verify this, we first use the classifier with three classes. Unknown damages should therefore be classified into the categories healthy, local and distributed damage. A distinction of the severity levels of the local and distributed damages cannot be tested due to the limited data. The three distributed damages are all defined as severe. In addition, there is a difficulty of quantifying the damage. A distinctive separation between light and heavy damage is non-trivial and subjective for each applicator. For these reasons, a classification of the severity is not provided.

The focus of the investigation is on the classification into the three presented subclasses. For the investigation, each damage class except *healthy* is withheld once in the training. The remaining damage phenomena are combined according to the subclasses and constitute the training data for the classifier. Subsequently, it is tested whether the withheld class can be classified into the correct subclass. The result is shown in Table 6. Here, the result for each withheld class is listed separately. Again, this test is performed with tenfold cross validation. The average accuracy and the accuracies of the best and worst iterations are listed.

**Table 6.** Classifier results for withheld classes with three subclasses.

| Unknown Class | $\mathrm{ACC_{mean}}$ | $\mathrm{ACC_{max}}$ | $\mathrm{ACC_{min}}$ |
|---|---|---|---|
| dist 0 | 0.03 | 0.29 | 0 |
| dist 1 | 0.99 | 1 | 0.98 |
| dist 2 | 1 | 1 | 0.99 |
| local 0 | 0.79 | 0.87 | 0.67 |
| local 1 | 0.43 | 0.77 | 0.23 |
| local 2 | 0.89 | 0.98 | 0.81 |

The result shows that classes *dist 1*, *dist 2* and *local 2* are classified almost perfectly. Class *local 0* shows moderate accuracy, whereas classes *dist 0* and *local 1* cannot be classified with sufficient accuracy. Class *dist 0* in particular performs very poorly, which is rather unexpected especially in comparison to the very good results of the other distributed damage classes.

A closer look at the misclassifications of withheld class *dist 0* reveals that most of the predictions belong to subclass *local*. In contrast to the classification of a healthy gearbox if damage is present, this does not represent a safety risk. In Figure 5, the CM of the best iteration of *local 1* is shown. Although most of the misclassifications are at the distributed damage and are therefore not critical, there are significantly more misclassifications at *healthy* than in the previous case. Therefore, although it yields better accuracy than the tests with *dist 0*, this could turn into a safety risk, and therefore the usage should be considered depending on the application.

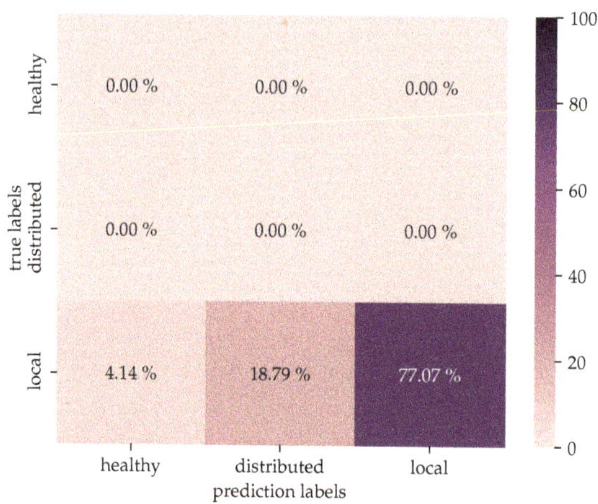

**Figure 5.** Confusion matrix for the classifier with three sub-classes tested on the withheld class *local 1*.

Because not all unknown damages can be classified in the defined subclasses, a second investigation differentiating only between healthy and damaged using the binary classifier is performed. Once again, every damage class is withheld one time and tested afterwards. The results are shown in Table 7. Like all other investigations, this is performed under tenfold cross validation, and therefore the average result as well as the best and the worst iterations are presented. Only class *local 0* shows slight difficulties, whereas all other classes exceed the average accuracies of over 0.95. For this reason, it can be derived that a reliable classification of unknown damage conditions into the classes healthy and damaged is achievable.

**Table 7.** Classifier results for withheld classes with two subclasses.

| Unknown Class | $ACC_{mean}$ | $ACC_{max}$ | $ACC_{min}$ |
|---|---|---|---|
| dist 0 | 0.96 | 0.99 | 0.90 |
| dist 1 | 0.98 | 1 | 0.97 |
| dist 2 | 1 | 1 | 1 |
| local 0 | 0.88 | 0.98 | 0.72 |
| local 1 | 0.98 | 1 | 0.97 |
| local 2 | 0.95 | 0.98 | 0.93 |

## 5. Conclusions

In this contribution, an investigation of the detection of various bearing damages is presented. Despite the measurement of the bearings in the compound of the two-stage planetary gearbox, which results in the superposition of vibration events, nonlinear transmisson paths, low-speed components and noise, the results show that a reliable detection of the bearing damage is achievable. A distinction can be made between six presented damage phenomena. For this purpose, vibration signals are collected by using an accelerometer. The vibration signals are evaluated by creating spectrograms, which are used by CNNs as input data.

In addition to the investigation of damage classification, complementary investigations regarding the robustness of the developed system are performed. These are intended to test the robustness of the system for field application as well as the expected variance in the damage cases. The results show that the classifier is robust for variations in the speeds within the trained speed ranges. Also in most cases, the classifier can achieve high accuracies in classifying unknown damages in the presented three sub-classes. However, due to difficulties with individual damage phenomena, the use of a binary classifier, which differentiates only between healthy and damaged, is recommended, especially for high-risk applications with an expected variance of damage phenomena to occur.

Since the classifier shows high robustness regarding unknown parameters, a transferability to planetary gearboxes with a different number of stages can be assumed. For this purpose, the network should probably be retrained. But due to missing access to equivalent data of a planetary gear with a different number of stages, a verification is not possible within this contribution. Therefore, the presented results are limited to the investigated two-stage planetary gearbox.

For future work, the classifier should be tested and optimized for bearing damages in the two-stage planetary gearbox installed in vehicles. Since the presented classifier can detect unknown damage phenomena, good transferability is assumed. Since unknown parameters, like additional noise or the influence of the load on the characteristics of the damage phenomena, are introduced in the field application, an adaptation of the classifier and additional filters are probably needed.

**Author Contributions:** Conceptualization, J.S.; methodology, J.S.; software, J.S. and L.J.B.; validation, J.S. and L.J.B.; formal analysis, J.S. and L.J.B.; investigation, J.S.; resources, J.S. and K.-L.K.; data curation, J.S.; writing—original draft preparation, J.S.; writing—review and editing, J.S., L.J.B., F.C. and K.-L.K.; visualization, J.S. and L.J.B.; supervision, K.-L.K.; project administration, K.-L.K.; funding acquisition, K.-L.K. All authors have read and agreed to the published version of the manuscript.

**Funding:** This work is funded by the German Federal Ministry of Economic Affairs and Climate Action project KISS (19I21015D).

**Data Availability Statement:** The data presented in this study are available on request from the corresponding author.

**Conflicts of Interest:** The authors declare no conflict of interest.

## Abbreviations

The following abbreviations are used in this manuscript:

| | |
|---|---|
| ACC | Accuracy |
| BAC | Balanced accuracy |
| BPFI | Ball pass-frequency inner race |
| BPFFI | Ball pass-frequency factor inner race |
| BPFO | Ball pass-frequency outer race |
| BPFFO | Ball pass-frequency factor outer race |
| BSF | Ball spin frequency |
| BSFF | Ball spin-frequency factor |
| CM | Confusion matrix |
| CNN | Convolutional neural network |
| FMD | Feature mode decomposition |
| FN | False negative |
| FP | False positive |
| FTF | Fundamental train frequency |
| FTFF | Fundamental train-frequency factor |
| TN | True negative |
| TP | True positive |

## References


1. Van Hecke, B.; Yoon, J.; He, D. Low speed bearing fault diagnosis using acoustic emission sensors. *Appl. Accoust.* **2016**, *105*, 35–44. [CrossRef]
2. Zhang, S.; Zhang, S.; Wang, B.; Habetler, T. Machine Learning and Deep Learning Algorithms for Bearing Fault Diagnostics—A Comprehensive Review. *IEEE Access* **2020**, *8*, 29857–29881. [CrossRef]
3. Segla, M.; Wang, S.; Wang, F. Bearing fault diagnosis with an improved high frequency resonance technique. In Proceedings of the IEEE 10th International Conference on Industrial Informatics (INDIN), Beijing, China, 25–27 July 2012; Volume 10, pp. 580–585.
4. Smith, W.; Randall, R. Rolling element bearing diagnostics using the Case Western Reserve University data: A benchmark study. *Mech. Syst. Signal Process.* **2015**, *64–65*, 100–131. [CrossRef]
5. Feng, Z.; Zhang, D.; Zuo, M. Adaptive Mode Decomposition Methods and Their Applications in Signal Analysis for Machinery Fault Diagnosis: A Review With Examples. *IEEE Access* **2017**, *5*, 24301–24331. [CrossRef]
6. Miao, Y.; Zhang, B.; Li, C.; Lin, J.; Zhang, D. Feature Mode Decomposition: New Decomposition Theory for Rotating Machinery Fault Diagnosis. *IEEE Trans. Ind. Electron.* **2023**, *70*, 1949–1960. [CrossRef]
7. Beering, A.; Döring, J.; Krieger, K.-L. Feature-based analysis of reproducible bearing damages based on a neural network. In Proceedings of the GMA/ITG-Fachtagung Sensoren und Messsysteme 2019, Nürnberg, Germany, 25–26 June 2019; Volume 20, pp. 451–456.
8. Yongbo, L.; Shubin, S.; Zhiliang, L.; Xihui, L. Review of local mean decomposition and its application in fault diagnosis of rotating machinery. *J. Syst. Eng. Electron.* **2019**, *30*, 799–814.
9. Lei, Y.; Lin, Y.; Zuo, M.; He, Z. Condition monitoring and fault diagnosis of planetary gearboxes: A review. *Measurement* **2014**, *48*, 292–305. [CrossRef]
10. Fischer, R.; Kücükay, F.; Jürgens, G.; Pollak, B. *Das Getriebebuch*, 2nd ed.; Springer: Wiesbaden, Germany, 2016; pp. 123–135.
11. Leaman, F.; Vicuna, C; Clausen, E. A Review of Gear Fault Diagnosis of Planetary Gearboxes Using Acoustic Emissions. *Acoust. Aust.* **2021**, *49*, 265–272. [CrossRef]
12. Zhang, X.; Wang, L.; Miao, Q. Fault diagnosis techniques for planetary gearboxes under variable conditions: A review. In Proceedings of the 2016 Prognostics and System Health Management Conference, Chengdu, China, 19–21 October 2016; pp. 1–11.
13. Mahgoun, H.; Ziani, R. Bearing Diagnostics Using Time-Frequency Filtering and EEMD. In *Rotating Machinery and Signal Processing*; Felkaoui, A., Chaari, F., Haddar, M., Eds.; Springer: Cham, Switzerland, 2019; pp. 44–55.
14. Tandon, N.; Choudhury, A. An analytical model for the prediction of the vibration response of rolling element bearings due to a localized defect. *J. Sound Vib.* **1997**, *205*, 275–292. [CrossRef]
15. Taye, M. Theoretical Understanding of Convolutional Neural Network: Concepts, Architectures, Applications, Future Directions. *Computation* **2023**, *11*, 52. [CrossRef]
16. Yamashita, R.; Nishio, M.; Do, R.; Togashi, K. Convolutional neural networks: An overview and application in radiology. *Insights Imaging* **2018**, *9*, 611–629. [CrossRef] [PubMed]
17. Dhillon, A.; Verma, G. Convolutional neural network: A review of models, methodologies and applications to object detection. *Prog. Artif. Intell.* **2020**, *9*, 85–112. [CrossRef]
18. Zhang, A.; Lipton, Z.; Li, M.; Smola, A. Dive into deep learning. *arXiv* **2021**, arXiv:2106.11342.
19. Guo, X.; Chen, L.; Shen, C. Hierarchical adaptive deep convolution neural network and its application to bearing fault diagnosis. *Measurement* **2016**, *93*, 490–502. [CrossRef]
20. Ajit, A.; Acharya, K.; Samanta, A. A Review of Convolutional Neural Networks. In Proceedings of the 2020 International Conference on Emerging Trends in Information Technology and Engineering (ic-ETITE), Vellore, India, 24–25 February 2020.
21. Qian, W.; Li, S.; Wang, J.; An, Z.; Jiang, X. An intelligent fault diagnosis framework for raw vibration signals: Adaptive overlapping convolutional neural network. *Meas. Sci. Technol.* **2018**, *29*, 095009. [CrossRef]
22. Kohavi, R. A Study of Cross-Validation and Bootstrap for Accuracy Estimation and Model Selection. In Proceedings of the 14th International Joint Conference on Artificial Intelligence, Montreal, QC, Canada, 20–25 August 1995; Volume 2, pp. 1137–1143.
23. Nandi, A.; Ahmed, H. Classification Algorithm Validation. In *Condition Monitoring with Vibration Signals: Compressive Sampling and Learning Algorithms for Rotating Machine*; Nandi, A., Ahmed, H., Eds.; John Wiley & Sons, Ltd.: New York, NY, USA, 2019; pp. 307–319.
24. Refaeilzadeh, P.; Tang, L.; Liu, H. Cross-Validation. In *Encyclopedia of Database Systems*; Liu, L., Ed.; Springer: Boston, MA, USA, 2009; pp. 532–538.
25. Tharwat, A. Classification assesment methods. *Appl. Comput. Inform.* **2021**, *17*, 168–192. [CrossRef]
26. Grandini, M.; Bagli, E.; Visani, G. Metrics for Multi-Class Classification: An Overview. *arXiv* **2020**, arXiv:2008.05756.
27. Brodersen, K.; Ong, C.; Stephan, K.; Buhmann, J. The Balanced Accuracy and Its Posterior Distribution. In Proceedings of the 2010 20th International Conference on Pattern Recognition, Istanbul, Turkey, 23–26 August 2010; pp. 3121–3124.

*Article*

# A Rapid and Inexpensive Method for Finding the Basic Parameters of Involute Helical Gears

**Stelian Alaci** [1,*], **Florina-Carmen Ciornei** [1], **Ionut-Cristian Romanu** [1], **Ioan Doroftei** [2,3], **Carmen Bujoreanu** [2] and **Ioan Tamaşag** [1]

[1] Mechanics and Technologies Department, "Stefan cel Mare" University of Suceava, 720229 Suceava, Romania; florina.ciornei@usm.ro (F.-C.C.); ionutromanucristian@usm.ro (I.-C.R.); ioan.tamasag@usm.ro (I.T.)

[2] Mechanical Engineering, Mechatronics and Robotics Department, "Gheorghe Asachi" Technical University, 700050 Iasi, Romania; idorofte@mail.tuiasi.ro (I.D.); carmen.bujoreanu@academic.tuiasi.ro (C.B.)

[3] Technical Sciences Academy of Romania, 26 Dacia Blvd, 030167 Bucharest, Romania

* Correspondence: stelian.alaci@usm.ro

**Abstract:** The paper proposes a rapid, straightforward, and inexpensive method for finding the basic parameters of helical gears with an involute profile. The basic parameters envisaged are the normal module, normal profile shift coefficient, and the helix angle. The proposed method uses balls introduced between the teeth and, thus, the contact with the measuring device surfaces is of the point type, and the centres of the balls are positioned symmetrically with respect to the measuring direction. The condition that the centre of the ball occupies an imposed position is mandatory. Additionally, there is the condition of the positions of the contact points between the balls and the flanks of the teeth. Two sets of balls of different sizes are necessary for a measurement. The conditions of the balls' positioning lead to a system of five unknowns. The methodology of solving the system is detailed and the method is exemplified for an actual helical gear. The new proposed method is based on the distance over pins but, using balls, presents the following advantages: It can be applied equally to all gears, regardless of the odd or even number of teeth. Furthermore, the dimension to be measured is singular compared to the dimension over pins when a maximum value must be found from several measurements.

**Keywords:** involute helical gear; inspection; module; profile shift coefficient; helix angle

**Citation:** Alaci, S.; Ciornei, F.-C.; Romanu, I.-C.; Doroftei, I.; Bujoreanu, C.; Tamaşag, I. A Rapid and Inexpensive Method for Finding the Basic Parameters of Involute Helical Gears. *Appl. Sci.* **2024**, *14*, 2043. https://doi.org/10.3390/app14052043

Academic Editor: Marek Krawczuk

Received: 31 January 2024
Revised: 26 February 2024
Accepted: 27 February 2024
Published: 29 February 2024

## 1. Introduction

Everyday engineering applications require new mechanisms which generate functions. To synthesise these mechanisms, the Machines and Mechanisms Theory must be applied. Thus, new dimensional constructive solutions of mechanisms, which, for a stipulated law of motion of the driving element, ensure a specified law of motion of the driven element [1], are designed. The problem can be solved in two ways: (a) the law of motion of the driven element is achieved with precision, and (b) the law of motion approximates, with a pre-stipulated accuracy, the imposed theoretical law. A more in-depth analysis shows that obtaining a solution which gives an imposed theoretical law of motion is an idealistic impossible task, mainly due to the backlash from the pairs of the mechanism, the manufacturing errors of the dimensions of the elements [2], and the deformations of the elements. Nevertheless, in a first approximation, it is accepted that all elements of the mechanism are perfectly rigid, the backlash from the kinematic pairs is zero, and the constructive dimensions of the elements are precisely machined. Under these simplifying hypotheses, one of the frequently met problems in engineering practice is finding the constructive solutions which allow, for the driven element, a motion identical to the motion of the driving element—these are the homokinetic mechanisms; Seherr-Thoss [3] and Dudita [4] present a series of constructive solutions of homokinetic mechanisms. An initial method to classify these mechanisms revolves round the relative position between the axes of the driven and driving element. In the broadest scenario when the driving and the driven axis are not co-planar, the tripodic transmissions are a solution of homokinetic coupling. Innocenti [5] presents a constructive solution where the tripodic coupling is achieved by pairs of point–surface type, while Qiu [6] presents a constructive solution of tripodic coupling with three curve–curve-type contacts. Notably for both methodologies, the mechanisms preserve homokinetic characteristics, even as the axes adjust their relative positioning over time. In the context of spherical mechanisms, an example of homokinetic mechanism is the Rzeppa coupling [7]. For planar mechanisms, where the input and output axes are parallel, the technical literature offers numerous constructive solutions for homokinetic mechanisms, with a prime example being the parallelogram mechanism. In instances where both axes retain fixed spatial positions, and the driving and driven element have rotation motion, the most convenient solution for achieving a homokinetic mechanism consists of a mechanism made of three elements only: the ground, the driving element, and the driven element. It is demonstrable that regardless of the position of the input and output axes, the transmission of motion between the mobile elements is made via a higher pair aligning with the overarching definition of cam mechanisms. This is a class 1 higher pair for the crossed axes and class 4 higher pair for the spherical mechanisms or plane parallel mechanisms [8]. This remark indicates the fact that the gear mechanisms are a special category of cam mechanisms. Additionally, one can attain broader relevance in the context of gear mechanisms by recalling the transmission ratio, generally defined as

$$i_{12} = \frac{\omega_1}{\omega_2} \tag{1}$$

where $\omega_1$ and $\omega_2$ are the angular velocities of the driven and driving axes, respectively. In the case of gear mechanisms, in most situations, they should ensure the transmission of motion with a rigorously constant ratio:

$$i_{12} = const. \tag{2}$$

The surfaces that give rise to the kinematic higher pair are typically referred to as "flanks". Specific criteria, as outlined in the pertinent technical literature [9,10], must be satisfied by these flanks to facilitate the transmission of rotational motion at a constant ratio.

For mechanisms with parallel axes, the flanks of the spur gears can be described as ruled surfaces, with their generatrixes lying parallel to the wheel axes. The whole gearing process can be studied in a section normal to the axes of the wheels (frontal plane). The intersection between the flanks and the plane normal to the axes of the wheels are named profiles. In this case, the condition of two toothed wheels to maintain a constant transmission ratio is derived from the fundamental law of gearing. This law dictates that for two profiles designed to mesh with a constant transmission ratio, the common normal at their point of contact should always pass through a fixed point [11,12]. This point known as the "pitch point " and lies along the centre line, dividing it in a ratio inversely proportional to the transmission ratio. Therefore, if there are stipulated the centers distance, the transmission ratio and one of the profiles, then, based on the fundamental law of gearing, the profile of the conjugate wheel can be found. From a practical point of view, an extremely pertinent problem is determining a profile curve that, by the gearing fundamental law, conducts to a curve of the same type of the conjugate profile. Therefore, the manufacturing technology of both teethed wheels is unified to a unique one and the costs of the transmission are thus substantially reduced. The single curve that satisfies this condition is the circle involute. This property of the involute makes possible the wide spread of spur gears with involute teeth in engineering applications. The parameters defining the involute teeth are used in all three stages: design, machining, and control. The last stage has a distinct importance since by finding the parameters which define the geometry of the teeth, one can estimate the manner of the fabrication process [13,14], ensure the accuracy requirements imposed on the gear during the design step, and thus certify the appropriate running of the assembly [15,16] from which it belongs. Concerning the control methods of the gears, these evolved from classical ones [17,18], when the over teeth and over pins/over balls dimensions were verified, to modern procedures [19–22]. One can mention a modern control method of gears which assimilates the helical gears with helical screws [23].

## 2. Materials and Methods

### 2.1. Aspects Concerning the Geometry of Spur Gears with Involute Teeth

Given that the geometric and kinematic evaluations of helical gears can ultimately be deduced from the analysis of a spur gear, we first consider the geometric and kinematic parameters of spur gears, before moving on to understanding those of a helical gear. From the above, it is plausible to leverage the fundamental law of gearing to determine a conjugate profile that corresponds to any given profile. From a manufacturing perspective, it is advantageous if both profiles are curves of the same type. Among all the planar curves, the involute of the circle is unique as its conjugate is of the same type. This curve is defined by a point attached to a mobile straight line that, without any slippage, rolls along an affixed circle, termed the base circle $C_b$, as shown in Figure 1.

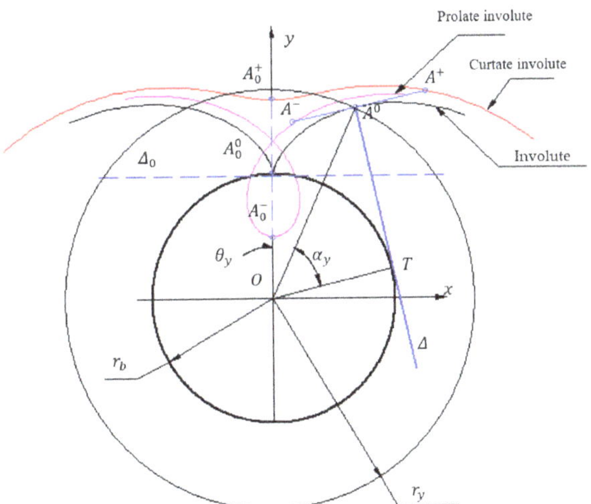


**Figure 1.** The involute of the circle: generation and properties.

In the initial position, the mobile straight line takes the position $\Delta_0$. Three fixed points on this straight line are considered: $A_0^0$ coincides to the point of tangency, $A_0^-$ is inside the base circle, and $A_0^+$. Allowing the straight line to roll without slipping over the base circle $C_b$, it occupies a current position $\Delta$ where the tangency with the base circle is in the point $T$. The three points fixed to the generating straight line take the current positions $A^0$, $A^-$, and $A^+$ respectively, describing during the motion of a normal involute, a prolate involute, and a curtate involute (Figure 1).

The condition of pure rolling stipulates that the length of the line segment $\overline{A_0T}$ should equal the arc length $\overset{\frown}{A_0^0T}$. Based on this remark, the polar parametric equations of the normal involute can be obtained:

$$\begin{cases} \rho_y = OA_0 = \dfrac{r_b}{\cos\alpha_y} \\ \theta_y = \angle A_0^0 O A^0 = \tan\alpha_y - \alpha_y \end{cases} \tag{3}$$

The function that appears in the definition of the $\theta_y$ angle is denoted $inv\alpha$ and is defined as

$$inv\alpha = \tan\alpha - \alpha, \ [\alpha]rad \tag{4}$$

and is frequently met in the theory of gearing. The $\alpha_y$ parameter from Equation (3) is the pressure angle of the normal involute on the base circle $C_y$ of radius $r_y$. Within the theory of involute gears, we can identify a circle by specifying the pressure angle of the normal involute profile. This profile is created using a base circle with an already established centre and radius.

Based on relations (3) and (4), the cartesian parametric equations of any of the three involutes can be obtained:

$$\begin{cases} x(\alpha) = \frac{r_b}{\cos\alpha}\sin(inv\alpha) + a \cdot \sin(tan\alpha) \\ y(\alpha) = \frac{r_b}{\cos\alpha}\cos(inv\alpha) + a \cdot \cos(tan\alpha) \end{cases} \tag{5}$$

where for $a < 0$ the prolate involute is obtained, for $a = 0$ the normal involute is obtained, and for $a > 0$ the curtate involute is obtained.

Another observation with important consequences concerns the fact that the length of the segment $TA_0$ is the curvature radius of the normal involute in the current point $A^0$.

$$\rho_c = length(TA_0) = r_b tan\alpha \tag{6}$$

As the tooth count of a gear wheel approaches infinity, it logically follows that the radius of the base circle expands towards infinity as well. Thus, a gear wheel evolves into a rack with an infinite number of teeth and the curvature radius of the rack profile also expands towards infinity. This implies that the profile of an involute gear rack takes the form of a straight line, as depicted in Figure 2. The base line of the rack $\Delta_0$ is the line alongside which the thickness of the tooth $s_0$ is equal to the space width $e_0$. Consequently, it corresponds to half the pitch of the rack (maintaining consistency across any section parallel to the direction of motion):

$$s_0 = e_0 = p_0/2 \tag{7}$$

with

$$p_0 = \pi m \tag{8}$$

where $m$ is a parameter with dimension of length and taking standard values. The reference straight line divides the tooth into two zones:

- The addendum, of height

$$h_{a0} = h_{a0}^* m, h_{a0}^* = 1 \leftrightarrow (ISO\leftrightarrow value); \tag{9}$$

- The dedendum. As can be noticed from the detail A, the tooth is cut on the inferior zone of a length

$$c_0 = c_0^* m, c_0^* = 0.25 (ISO\, value) \tag{10}$$

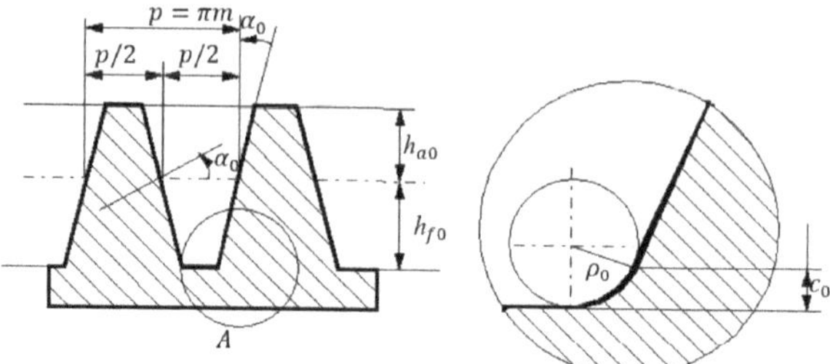

**Figure 2.** The basic rack with straight teeth.

The reference height of the dedendum becomes

$$h_{f0} = h_{f0}^* m = (h_{a0}^* + c_0^*)m \tag{11}$$

For a complete definition of the geometry of the rack, additional specifications must be stipulated:

- The pressure angle of the straight profile:

$$\alpha_0 = 20°; \tag{12}$$

- The filleting radius $\rho_0$ at the root of the tooth:

$$\rho_0 = \frac{c_0^*}{1 - sin\alpha_0}m \cong 0.38m \tag{13}$$

As a conclusion, the profile of the basic rack is defined by the module $m$, the pressure angle $\alpha_0$, the reference addendum coefficient $h_{a0}^* = 1$, and the coefficient of radial backlash $c_0^* = 0.25$.

The gear with involute teeth is defined using the basic rack. The definition parameters require adherence to two specific conditions:

- The gearing is accomplished without backlash between the flanks;
- The gearing is accomplished with standard radial backlash $c_0$

The two conditions are sufficient for the complete definition of the gear. The centroids of the relative motion between the wheel and the base rack are represented (Figure 3) by the pitch circle of the gear $c_d$ of diameter

$$d = mz, \tag{14}$$

and the rolling straight line of the rack $\Delta_w$ parallel to the reference line $\Delta_0$, but at a distance from this, with the quantity $X$ named profile shift and expressed as

$$X = mx \tag{15}$$

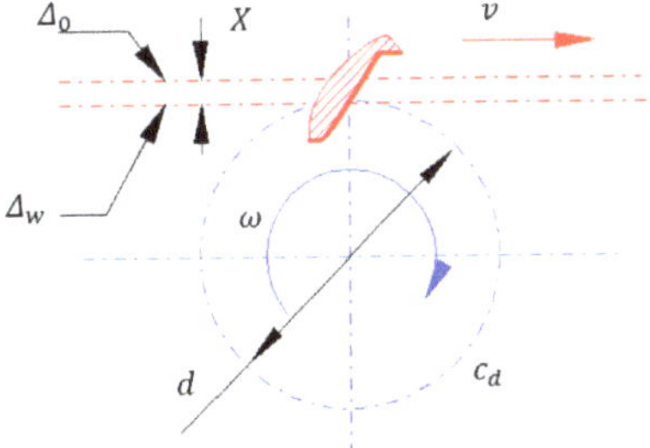

**Figure 3.** The pitch straight line of the rack rolls over the pitch circle of the gear; the profile shift definition.

The profile shift coefficient $x$ of the gear is considered:
$x = 0$ when the straight line $\Delta_w$ is tangent to the pitch circle;
$x < 0$ when $\Delta_w$ intersects the pitch circle;
$x > 0$ when $\Delta_w$ is external to the pitch circle.

In conclusion, with stipulated parameters for the shape and dimensions of the basic rack, the toothed wheel is fully characterized by the next three parameters:

1. The number of teeth $z$, which describes the dimension of the wheel;
2. The module $m$, which describes the dimension of the tooth;

3.   The profile shift coefficient $x$, which describes the shape of the tooth.

A precise method for manufacturing the involute gears is gear generation, wherein the resultant gear is conceived as the envelope of a distinctive series of positions of a toothed cutting tool. This procedure requires a relative motion between the gear blank and the tool mimicking the motion that would otherwise be present between the tool and the final, machined gear. One manufacturing technique uses a tool which mirrors the profile of the generating rack (conjugate to the basic rack).

To highlight the impact of the number of teeth and of the profile shift coefficient on gear machining, a simulation program was developed. This simulation experiment incorporated two different configurations of tooth count, $z = 5$ and $z = 10$, along with three distinct values attributed to the profile shift coefficient. Figure 4 shows the results of the simulation. It is evident that negative values of the profile shift coefficient led to an undercutting phenomenon which involves the removal of tooth material within the dedendum zone. Conversely, positively large values result in overcutting of the tooth. Both phenomena are more pronounced when the tooth count of the gear wheel $z$ is smaller.

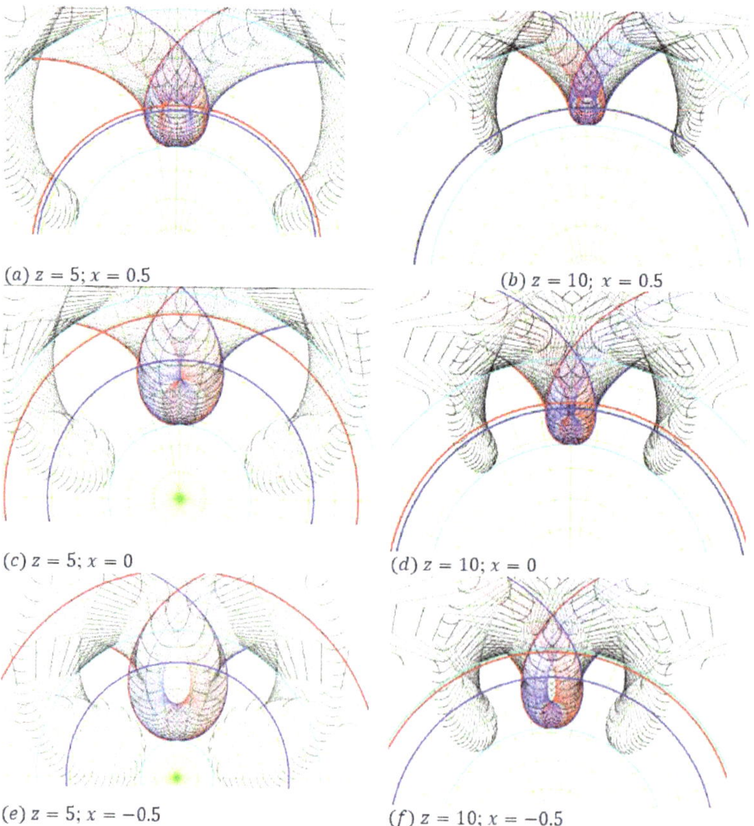


**Figure 4.** The effect of number of teeth $z$ and of the profile shift coefficient $x$ upon the shape of the wheel machined with tools of rack type.

Concerning the adoption of the profile shift coefficient of a gearing, essential observations are as follows:

- The first remark is related to the effect of the profile shift coefficient upon the centre distance; in order to achieve a standard centre distance, gear correction must be performed;
- The second observation refers to the gear synthesis methodology; to ensure satisfactory kinematic and dynamic behaviour, a gear mechanism needs to conform to all stipulated conditions for smooth operation. Such criteria can be defined by a condition formulated as follows:

$$F_k(m, z_1, z_2, x_1, x_2) \geq 0, \quad k = 1 \div n \tag{16}$$

where $n$ is the number of imposed criteria. For the imposed module, the relation (16) becomes

$$f_k(z_1, z_2, x_1, x_2) \geq 0 \tag{17}$$

If we consider only the equality from Equation (17) and treat the profile shift coefficients $x_1$ and $x_2$ while $z_1$ and $z_2$ are regarded as parameters, the equation of a curve is obtained:

$$\phi_k(x_1, x_2) \geq 0 \tag{18}$$

This curve divides the plane $x_1 x_2$ into two regions: one where the criterion is satisfied, and another where it is not. For a gearing with a stipulated number of teeth, the intersection of all regions where all the $n$ conditions are verified will represent, in the $x_1 x_2$ plane, a closed domain [24], where all the imposed criteria of correct running will be satisfied. The boundary of this domain is named the "locking contour". Knowing the locking contour allows, for a given pair of teeth, the correct and operative choice of the profile shift coefficients.

### 2.2. Helical Gears

In order to define the profile of a helical gear, a base cylinder is set on a plane $(P)$, as shown in Figure 5. In the $(P)$ plane, $(\Delta)$ is considered the straight line that makes the angle $(\beta_b)$ with the generatrix of the cylinder tangent to the plane. By rolling the plane $(P)$ without slipping, a ruled surface is generated which can be used as a tooth flank. During the process of flank creation, the points of the straight line $(\Delta)$ will sit on the base cylinder as a helix with constant pitch $(E_b)$, named the base helix. Each point of the straight line $(\Delta)$ describes an involute generated using the circles of radius $(r_b)$ (the radius of the base cylinder). As an example, in the foreground plane, the involute $(e)$ is generated and in the background plane, the involute $(e')$ is generated. All involute curves commence from the base cylinder and, due to their origin at the base helix, preserve a uniform angular discrepancy across the length of the cylinder's generatrix.

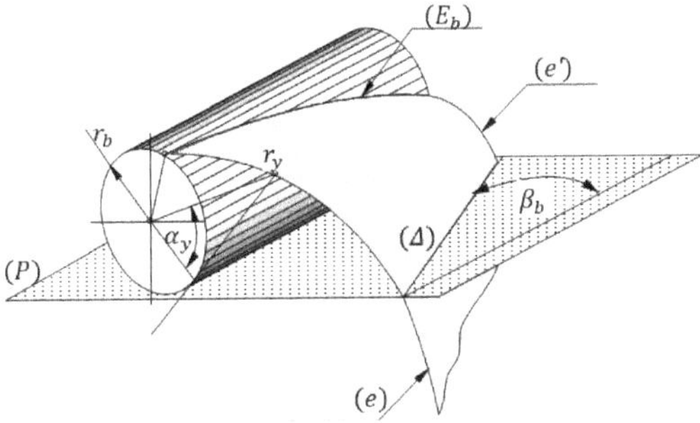


**Figure 5.** Defining the flank for a helical spur gear with involute teeth.

The generation process of the flank's surface yields the conclusion that it constitutes a helical involute ruled surface. The intersection of the flank with an arbitrary cylinder of radius $r_y$ coaxial to the base cylinder is a helix $(E_y)$. Based on Figure 6, where the unfolded cylinders are also shown, the inclination angle $\beta_y$ of the helix on the cylinder of radius $r_y$ is found. It must be noted that both helices share the same axial pitch.

$$\left. \begin{array}{l} tan(\beta_y) = \frac{2\pi r_y}{p_z} \\ tan(\beta_b) = \frac{2\pi r_b}{p_z} \end{array} \right\} \Rightarrow tan(\beta_y) = tan(\beta_b)\frac{r_y}{r_b} \tag{19}$$

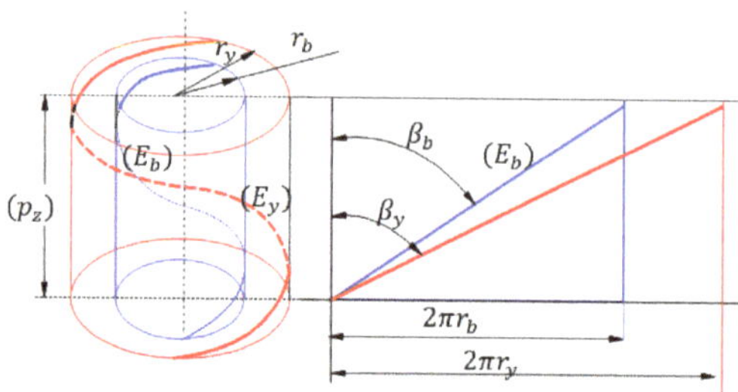

**Figure 6.** Establishing the angle of the tooth helix on an arbitrary cylinder.

We introduce the convention that the parameters from a frontal section (normal to the axis of the wheel) are denoted with the index "$t$". Thus, the relation (19) becomes

$$tan(\beta_y) = \frac{tan(\beta_y)}{cos(\alpha_{yt})} \tag{20}$$

where $\alpha_{yt}$ represents the pressure angle of the involute $(e)$ on the circle of radius $r_y$, as in Figure 6. Given that the gearing interaction between the profiles in any section perpendicular to the axes of the wheels mirrors that of two spur gears, it can be asserted that the rack's profile will be linear across all frontal sections. Additionally, with helical teeth, the interaction between the teeth of the wheel and the rack occurs along a straight line $(\Delta)$. It is noteworthy that this line no longer aligns perpendicularly to the direction of displacement. As a result, it becomes apparent that the flank of the helical rack is planar, with the orientation of the teeth not being perpendicular to the direction of displacement. Instead, the teeth's direction forms an angle $\beta_0$ with the direction of displacement. In a section perpendicular to the tooth, the size and the design of the reference helical rack must have standard dimensions, identical to that observed in a rack with spur teeth. Figure 7 shows the basic rack with helical teeth.

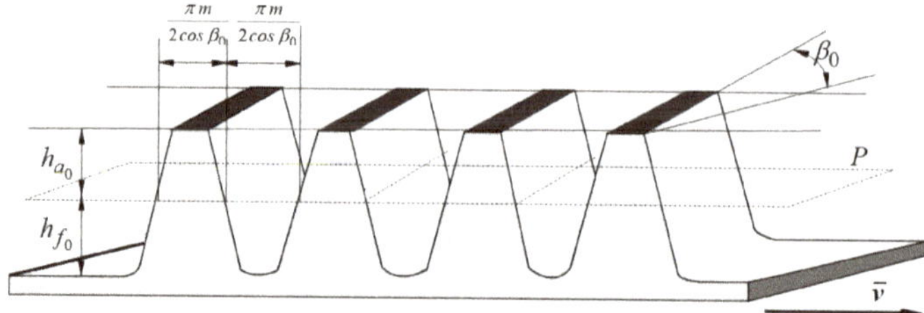

**Figure 7.** The basic rack with helical teeth.

The base plane $P$ of the rack is presented in Figure 7. The intersection between the reference plane to any frontal plane is a straight line along which the tooth thickness is identical to the space width. With respect to the reference plane, the following parameters are defined:

- The addendum of the tooth $h_{a_0}$;

- The dedendum of the tooth $h_{f_0}$.

In the frontal plane, the frontal pitch $p_t$, and $h_{a_0}$ and $h_{f_0}$ are defined as functions of the frontal module $m_t$ using relations similar to the ones used for spur gears:

$$p_t = \pi m_t \tag{21}$$

$$h_{a_0 t} = h^*_{a_0 t} m_t \tag{22}$$

$$h_{f_0 t} = (h^*_{a_0 t} + c^*_{0t}) m_t \tag{23}$$

In order to apply the relations (21)–(23), the relations between the parameters from the frontal section $h^*_{a_0 t}$, $c^*_{0t}$, and $m_t$ and the known (standard) parameters from the normal section $h^*_{a_0} = 1$ and $c^*_0 = 0.25\,m$ must be found. Figure 8 shows a section with the base plane through the basic rack with helical teeth, a section normal to the tooth denoted $n$-$n$, and another frontal one $t$-$t$ parallel to the direction of motion of the basic rack. The geometric parameters in the frontal section are indexed with "$t$", along with the convention that the parameters from the normal section are not indexed. It should be noted that in other works the "$n$" index may be used instead. From the right triangle QMP, one can write

$$PM = \frac{MQ}{cos(\beta_0)} \Rightarrow p_t = \frac{p}{cos(\beta_0)} \tag{24}$$

and from here, the relation between the normal $m$ and frontal $m_t$ module results in

$$m_t = \frac{m}{cos(\beta_0)} \tag{25}$$

The addendum $h_{a_0}$ and the dedendum $h_{f_0}$ of the tooth appear in actual size, both in the normal and frontal sections

$$h_{a_0} = h^*_{a_0 t} m_t = h^*_{a_0} m \tag{26}$$

where from

$$h^*_{a_0 t} = h^*_{a_0} \frac{m}{m_t} = h^*_{a_0} cos(\beta_0) \tag{27}$$

it results

$$h_{f_0} = (h^*_{a_0 t} + c^*_{0t}) m_t = (h^*_{a_0} + c^*_0) m \tag{28}$$

where, considering Equation (27), one obtains

$$c^*_{0t} m_t = c^*_0 m \tag{29}$$

and from here, the following is obtained:

$$c^*_{0t} = c^*_0 cos\beta_0 \tag{30}$$

To find the pressure angle of the profile in the frontal plane, the triangles $ABC$ and $A_t B_t C_t$ are used, where

$$A_t B_t = AB = h_{a_0}; \; BC = B_t C_t cos\beta_0; \; BC = h_{a_0} tan\alpha_0; \; B_t C_t = h_{a_0} tan\alpha_{0t} \tag{31}$$

From relation (31), one obtains

$$tan(\alpha_{0t}) = \frac{tan(\alpha_0)}{cos(\beta_0)} \tag{32}$$

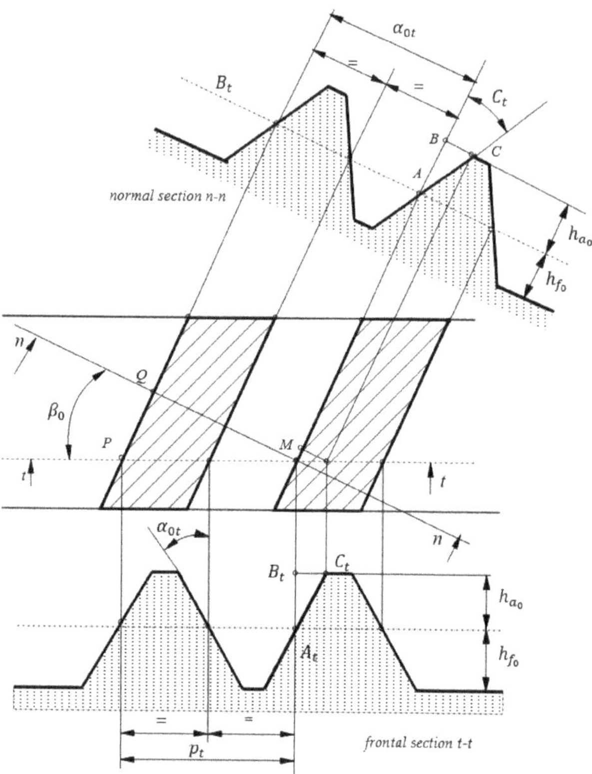


**Figure 8.** Sections through basic rack with helical teeth: a section with the base plane (hatched), a section normal to the tooth (*n-n*), and a frontal section (*t-t*).

The relations (26)–(32) fully define the profile of the basic rack in the frontal section. Therefore, the wheel can be defined in the frontal section using the frontal profile of the base rack. The conditions for definition are the same as for the spur teeth:

- Gearing without backlash between flanks;
- Gearing with standard radial backlash;
- The shift of the profile X is the same in both sections.

$$X = x_t m_t = xm, \Rightarrow x_t = x\frac{m}{m_t} = x\cos\beta_0 \tag{33}$$

The frontal pitch $p_t$ is

$$p_t = \pi m_t = \pi\frac{m}{\cos\beta_0} \tag{34}$$

and the pitch diameter $d$ is

$$d = zm_t = \frac{z}{\cos\beta_0}m \tag{35}$$

The addendum of the tooth $h_a$ is

$$h_a = (h^*_{a_0t} + x_t)m_t = (h^*_{a_0} + x)m \tag{36}$$

The dedendum of the tooth $h_f$ is

$$h_f = (h^*_{a_0t} + c^*_{0t} - x_t)m_t = (h^*_{a_0} + c^*_0 - x)m \tag{37}$$

The addendum diameter $d_a$ is

$$d_a = d + 2h_a = \left(z + 2h^*_{a_0t} + 2x_t\right)m_t = \left(\frac{z}{cos\beta_0} + 2h^*_{a_0} + 2x\right)m \qquad (38)$$

The pressure angle of the frontal profile on the pitch circle $\alpha_t$ is

$$\alpha_t = \alpha_{0t} = \text{atan}\frac{tan\alpha_0}{cos\beta_0} \qquad (39)$$

The base diameter $d_b$ is

$$d_b = dcos\alpha_{0t} = m_t z cos\alpha_{0t} = \frac{mz}{cos\beta_0}cos\alpha_{0t} \qquad (40)$$

The angle of inclination of the helix $\beta$ of the flank on the pitch cylinder is

$$\beta = \beta_0 \qquad (41)$$

The relation (20) -the angle of the flank's helix in the base cylinder $\beta_b$ is applied for the case of the base circle and is

$$\beta_b = \text{atan}[tan\beta_0 cos\alpha_{0t}] \qquad (42)$$

*2.3. Method and Device for Finding the Characteristic Parameters of the Helical Gears with Involute Teeth*

2.3.1. The Procedure for Finding the Characteristic Parameters of a Helical Gear with Involute Teeth

In a recent work, the authors proposed a quick and cost-effective technique, along with the associated device, to accurately determine the three geometric characteristics of an involute toothed spur gear: the number of teeth $z$, the module $m$, and the profile shift coefficient $x$. The method is based on the manner of precision control of the spur gear with involute teeth using the measurement over pins. Two main disadvantages are obvious for the method of over pins measurement:

- The mechanical system formed by the measuring instrument, the two pins, and the measured wheel is not steady, and the distance to be measured is accepted as the maximum value of the distance between the two pins;
- The application of the method varies depending on whether the number of teeth is odd or even.

The method proposed in [25] eliminates these drawbacks by employing a number of three or four pins, placed symmetrically with respect to the measuring direction, Figure 9a (even teeth number) and Figure 9b (odd teeth number).

As a fundamental premise, the method involves experimentally determining the dimension over pins by utilizing a specifically designed device and expressing this dimension in terms of the gear's three primary geometric features. Finally, an equation of two unknowns, the module $m$ and the profile shift coefficient $x$, is obtained:

$$inv\left\{acos\left[\frac{mzcos\alpha_0}{2(D-d_R)}(cos\psi_{up} + cos\psi_{down})\right]\right\} = inv\alpha_0 - \frac{\pi}{2z} + \frac{2x}{z}tan\alpha_0 + \frac{d_R}{mz\alpha_0} \qquad (43)$$

Therefore, two equations are necessary to find the two parameters $m$ and $x$. To overcome this, it is essential to note that the dimensions of the wheel do not undergo significant change for the typical values of the profile shift coefficient:

$$-0.5 < x < 0.5 \qquad (44)$$

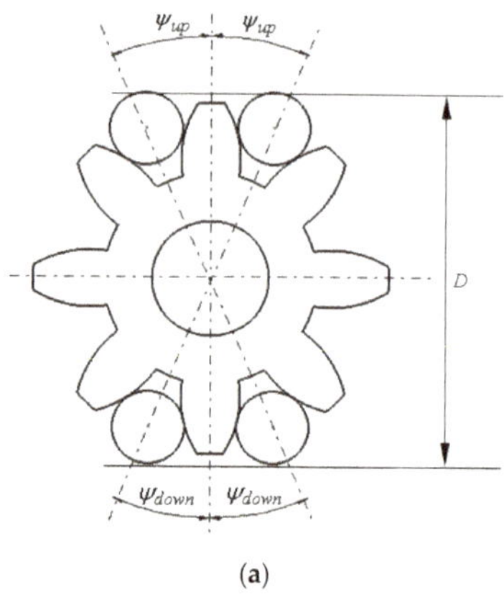
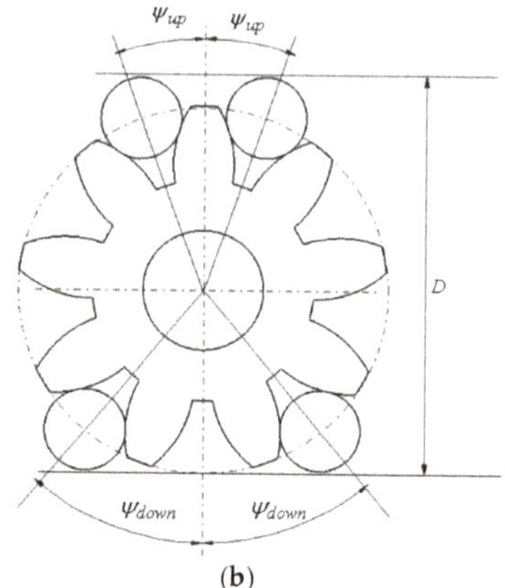



**Figure 9.** The proposed method for finding the characteristic parameters of a spur gear with involute teeth: (**a**) gear with even number of teeth, when the direction of measurement passes through two teeth or two spaces; (**b**) gear with odd number of teeth, when the direction of measurement passes through a tooth and a space.

So, by replacing $x = 0$ in Equation (43), the following equation of the unknown module $m$ is obtained:

$$inv\left\{acos\left[\frac{mzcos\alpha_0}{2(D-d_R)}(cos\psi_{up} + cos\psi_{down})\right]\right\} = inv\alpha_0 - \frac{\pi}{2z} + \frac{d_R}{mz\alpha_0} \tag{45}$$

This equation can be solved simply. Next, knowing that the module $m$ takes standard values, the actual module of the gear is chosen as the standard value $m_{std}$, closest to the solution of Equation (45). Now, the adopted standard value of the module $m_{std}$ is replaced in Equation (43) and an equation from which the profile shift coefficient $x$ is found is obtained:

$$inv\left\{acos\left[\frac{m_{std}zcos\alpha_0}{2(D-d_R)}(cos\psi_{up} + cos\psi_{down})\right]\right\} = inv\alpha_0 - \frac{\pi}{2z} + \frac{2x}{z}tan\alpha_0 + \frac{d_R}{m_{std}z\alpha_0} \tag{46}$$

### 2.3.2. Alternative Method for the Helical Gears

The methodology outlined above is not applicable to involute helical gears. The main reason for this is the difficulty of employing pins or rollers with this type of gear. As illustrated in Figure 10, the axes of the two pins making contact with the tooth flanks are not coplanar, making this approach impractical.

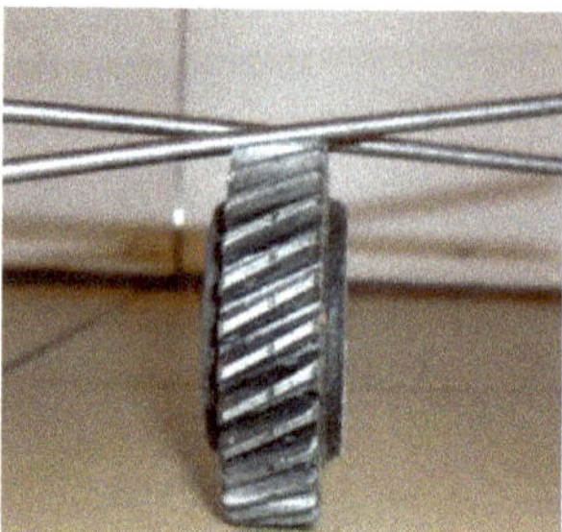

**Figure 10.** The axes of the pins are not coplanar.

This can be overcome by using ball bearings spheres, keeping in mind that these balls should be in contact with the flanks of the teeth and with another frontal surface. The motivation behind this approach is to ensure that all the sphere centres are situated within the same frontal plane, as demonstrated in Figure 11.

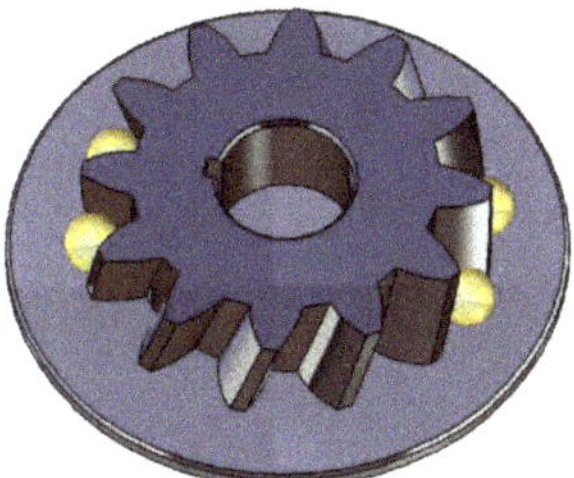

**Figure 11.** Schematics for the device used for involute helical gears.

The geometry of the wheel is characterised by the number of teeth $z$, the module $m$ in the normal section, the normal profile shift coefficient $x$, and the helix angle of the teeth of the base rack $\beta_0$. The gearing is studied in the frontal section, and the profile is also an involute, characterised by the number of teeth $z$, the frontal module $m_t$, and the frontal profile shift coefficient $x_t$.

Considering the relation (25) and the fact that the angle $\beta_0$ may take any value, it results that in the frontal section, the value of the module is not standard and the methodology applied for the spur gears is not effective. Therefore, in this case, the conditions imposed upon the coordinates of the contact points are used.

In Figure 12, a frontal section is presented, which contains the centre of a ball of radius $r_B$ which makes contact with the boundary flanks at points $Y$ and $Y$. The normal at the contact point is tangent to the base circle in the $T_y$ point. The $Y$ points are placed on a circle characterised by the pressure angle $\alpha_{yt}$ of the involute generated using the base circle of radius $r_b$. The pitch circle passes through points $D$.

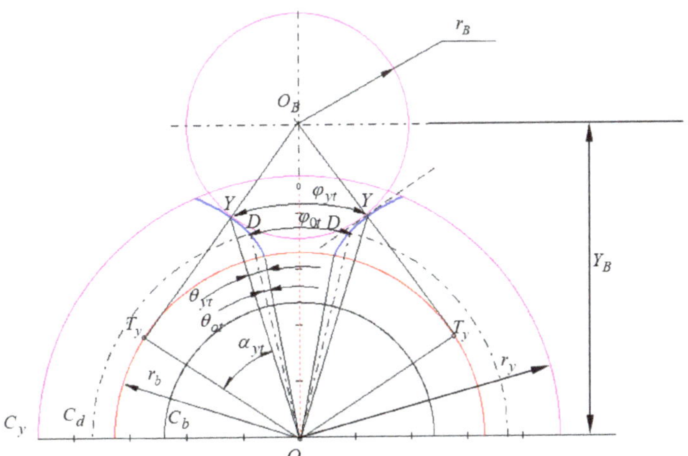

**Figure 12.** Finding the coordinates of the contact points.

Applying the involute properties, one can write

$$arcY'Y'' = r_y \varphi_{yt} = \frac{1}{2}\frac{mz\cos\alpha_{0t}}{\cos\alpha_{yt}}\left[\varphi_{0t} + 2(\theta_{yt} - \theta_{0t})\right] = \frac{1}{2}\frac{mz\cos\alpha_{0t}}{\cos\alpha_{yt}}\left[\frac{e_d}{d/2} + 2(inv\alpha_{yt} - inv\alpha_{0t})\right]$$
$$= \frac{mz\cos\alpha_{0t}}{\cos\alpha_{yt}}\left[\frac{\frac{\pi}{2}-2x_t\tan\alpha_{0t}}{z} + (inv\alpha_{yt} - inv\alpha_{0t})\right] = \frac{mz\cos\alpha_{0t}}{\cos\alpha_{yt}}\left[\frac{\pi}{2z} - \frac{2x_t\tan\alpha_{0t}}{z} + (inv\alpha_{yt} - inv\alpha_{0t})\right] \tag{47}$$

where $\varphi_y$ represents the arc of the space width on the circle $C_y$ and $\varphi_{0t}$ is the arc of the space width on the pitch circle. From the above relation, one can obtain

$$\varphi_{yt} = \frac{arcYY}{r_y} = \frac{\frac{mz\cos\alpha_{0t}}{\cos\alpha_{yt}}\left[\frac{\pi}{2z} - \frac{2x_t\tan\alpha_{0t}}{z} + (inv\alpha_{yt} - inv\alpha_{0t})\right]}{\frac{1}{2}\frac{mz\cdot\cos\alpha_{0t}}{\cos\alpha_{yt}}} = 2\left[\frac{\pi}{2z} - \frac{2x_t\tan\alpha_{0t}}{z} + (inv\alpha_{yt} - inv\alpha_{0t})\right] \tag{48}$$

Next, one can obtain

$$T_Y O T_Y = \varphi_{yt} + 2\alpha_{yt} = 2\left[\frac{\pi}{2z} - \frac{2x_t\tan\alpha_{0t}}{z} + (inv\alpha_{yt} - inv\alpha_{0t})\right] + 2\alpha_{yt}$$
$$= 2\left[\frac{\pi}{2z} - \frac{2x_t\tan\alpha_{0t}}{z} + (inv\alpha_{yt} + \alpha_{yt} - inv\alpha_{0t})\right] = 2\left[\frac{\pi}{2z} - \frac{2x_t\tan\alpha_{0t}}{z} + \tan\alpha_{yt} - inv\alpha_{0t}\right] \tag{49}$$

The angles $T_Y O_B T_Y$ and $T_Y O T_Y$ are supplementary:

$$T_y O_B T_y = \pi - 2\left[\frac{\pi}{2z} - \frac{2x_t\tan\alpha_{0t}}{z} + \tan\alpha_{yt} - inv\alpha_{0t}\right] \tag{50}$$

The length of the cord $YY$ can be expressed in two ways:

$$r_B\sin\left[\frac{1}{2}\left(\pi - 2\left[\frac{\pi}{2z} - \frac{2x_t\tan\alpha_{0t}}{z} + \tan\alpha_{yt} - inv\alpha_{0t}\right]\right)\right]$$
$$= r_y\sin\left[\frac{1}{2}\left(2\left[\frac{\pi}{2z} - \frac{2x_t\tan\alpha_{0t}}{z} + (inv\alpha_{yt} - inv\alpha_{0t})\right]\right)\right] \tag{51}$$

And from here, the following equation is obtained:

$$r_B\cos\left(\frac{\pi}{2z} - \frac{2x_t\tan\alpha_{0t}}{z} + \tan\alpha_{yt} - inv\alpha_{0t}\right) = \frac{1}{2}mz\frac{\cos\alpha_{0t}}{\cos\alpha_{yt}}\sin\left[\frac{\pi}{2z} - \frac{2x_t\tan\alpha_{0t}}{z} + (inv\alpha_{yt} - inv\alpha_{0t})\right] \tag{52}$$

Additionally, the radius of the centre of the ball can be found:

$$Y_B = r_B\cos\left[\frac{1}{2}\left(\pi - 2\left[\frac{\pi}{2z} - \frac{2x_t\tan\alpha_{0t}}{z} + \tan\alpha_{yt} - inv\alpha_{0t}\right]\right)\right]$$
$$+ r_y\cos\left[\frac{1}{2}\left(2\left[\frac{\pi}{2z} - \frac{2x_t\tan\alpha_{0t}}{z} + (inv\alpha_{yt} - inv\alpha_{0t})\right]\right)\right] \tag{53}$$

From here, the following is obtained:

$$Y_B = r_B\sin\left(\frac{\pi}{2z} - \frac{2x_t\tan\alpha_{0t}}{z} + \tan\alpha_{yt} - inv\alpha_{0t}\right) + \frac{1}{2}mz\frac{\cos\alpha_{0t}}{\cos\alpha_{yt}}\cos\left[\frac{\pi}{2z} - \frac{2x_t\tan\alpha_{0t}}{z} + (inv\alpha_{yt} - inv\alpha_{0t})\right] \tag{54}$$

Figure 13 shows that the measured dimension over balls $D$ using the proposed device can be expressed as

$$D = Y_B\cos\psi_{up} + Y_B\cos\psi_{down} + 2r_B \tag{55}$$

where $Y_B$ is the radius of the circle where the centres of the balls (of radii $r_B$) are positioned in the space between teeth, contacting both flanks of the teeth. As previously stated, the balls are positioned symmetrically with respect to the diameter parallel to the direction of measurement. From relation (55), the following is obtained:

$$Y_B = \frac{D - 2r_B}{\cos\psi_{up} + \cos\psi_{down}} \tag{56}$$

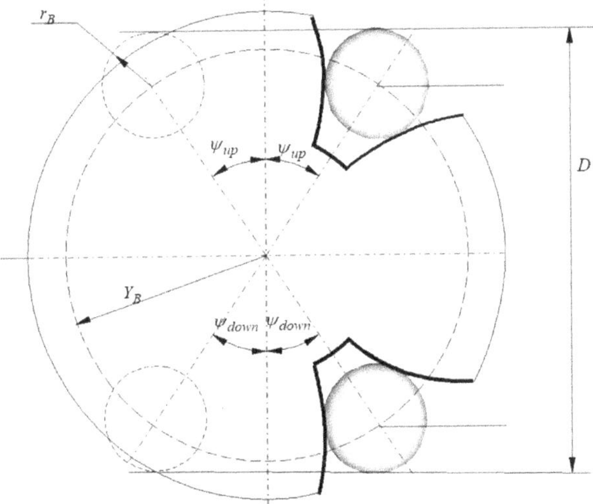

**Figure 13.** The dimension over balls for the proposed method.

The relations (54) and (56) result in the equation:

$$r_B sin\left(\frac{\pi}{2z} - \frac{2x_t tan\alpha_{0t}}{z} + tan\alpha_{yt} - inv\alpha_{0t}\right) + \frac{1}{2}mz\frac{cos\alpha_{0t}}{cos\alpha_{yt}}cos\left[\frac{\pi}{2z} - \frac{2x_t tan\alpha_{0t}}{z} + (inv\alpha_{yt} - inv\alpha_{0t})\right] = \frac{D - 2r_B}{cos\psi_{up} + cos\psi_{down}} \tag{57}$$

Now, considering the Equations (52) and (57), the following system is obtained:

$$\begin{cases} r_B cos\left(\frac{\pi}{2z} - \frac{2x_t tan\alpha_{0t}}{z} + tan\alpha_{yt} - inv\alpha_{0t}\right) = \frac{1}{2}mz\frac{cos\alpha_{0t}}{cos\alpha_{yt}}sin\left[\frac{\pi}{2z} - \frac{2x_t tan\alpha_{0t}}{z} + (inv\alpha_{yt} - inv\alpha_{0t})\right] \\ r_B sin\left(\frac{\pi}{2z} - \frac{2x_t tan\alpha_{0t}}{z} + tan\alpha_{yt} - inv\alpha_{0t}\right) + \frac{1}{2}mz\frac{cos\alpha_{0t}}{cos\alpha_{yt}}cos\left[\frac{\pi}{2z} - \frac{2x_t tan\alpha_{0t}}{z} + (inv\alpha_{yt} - inv\alpha_{0t})\right] = Y_B \end{cases} \tag{58}$$

The unknowns of the system are

- the frontal module $m_t$;
- the frontal pressure angle $\alpha_{yt}$ on the circle passing through the contact points;
- the frontal profile shift coefficient of the gear $x_t$;
- The frontal pressure angle $\alpha_{0t}$ on the pitch cylinder of the gear.

Considering that the method is applied using two sets of balls of radii $r_{B1}$ and $r_{B2}$ positioned at the angles $\psi_{down1}$, $\psi_{up1}$, $\psi_{down2}$, and $\psi_{up2}$, respectively, the system (58) written for the two cases leads to a system of four equations which, after some calculations, takes the form

$$\begin{cases} \frac{mz}{2Y_{B1}}cos\alpha_{0t} - cos\left[asin\left(\frac{r_{B1}}{Y_{B1}}cos\alpha_{y1t}\right) + \alpha_{y1t}\right] = 0 \\ r_{B1}cos\alpha_{y1t} - Y_{B1}\left[\frac{\pi}{2z} - \frac{2x_t tan\alpha_{0t}}{z} + inv\alpha_{y1t} - inv\alpha_{0t}\right] = 0 \\ \frac{mz}{2Y_{B2}}cos\alpha_{0t} - cos\left[asin\left(\frac{r_{B2}}{Y_{B2}}cos\alpha_{y1t}\right) + \alpha_{y2t}\right] = 0 \\ r_{B2}cos\alpha_{y2t} - Y_{B2}\left[\frac{\pi}{2z} - \frac{2x_t tan\alpha_{0t}}{z} + inv\alpha_{y2t} - inv\alpha_{0t}\right] = 0 \end{cases} \tag{59}$$

A system of two transcendental equations (from the first and third equation of system (59) and, respectively, from the second and fourth equation) of unknowns $\alpha_{y1t}$ and $\alpha_{y2t}$ results in

$$\begin{cases} Y_{B1}cos\left[asin\left(\frac{r_{B1}}{Y_{B1}}cos\alpha_{y1t}\right) + \alpha_{y1t}\right] - Y_{B2}cos\left[asin\left(\frac{r_{B2}}{Y_{B2}}cos\alpha_{y2t}\right) + \alpha_{y2t}\right] = 0 \\ inv\alpha_{y1t} - asin\left(\frac{r_{B1}}{Y_{B1}}cos\alpha_{y1t}\right) - inv\alpha_{y2t} + asin\left(\frac{r_{B2}}{Y_{B2}}cos\alpha_{y2t}\right) = 0 \end{cases} \tag{60}$$

The two unknowns $\alpha_{y1t}$ and $\alpha_{y2t}$ are found via a numerical method. After the two pressure angles are found, two equations of system (59) are considered (either the first two, or the last two equations). We consider here the first two equations of system (59) and with the relations (25), (33), and (39) between the geometrical parameters from the normal and frontal section; the next system of equations results in

$$\begin{cases} \dfrac{m_t z}{2Y_{B1}}\cos\alpha_{0t} - \cos\left[asin\left(\dfrac{r_{B1}}{Y_{B1}}\cos\alpha_{y1t}\right) + \alpha_{y1t}\right] = 0] \\ r_{B1}\cos\alpha_{y1t} - Y_{B1}\left[\dfrac{\pi}{2z} - \dfrac{2x_t \tan\alpha_{0t}}{z} + inv\alpha_{y1t} - inv\alpha_{0t}\right] = 0 \end{cases}$$

(61)

Now, we recall Equations (25), (39), and (54) and have

$$\begin{cases} \dfrac{m\cos\beta_0}{2Y_{B1}}z\cos\left(atan\dfrac{\tan\alpha_0}{\cos\beta_0}\right) - \cos\left[asin\left(\dfrac{r_{B1}}{Y_{B1}}\cos\alpha_{y1t}\right) + \alpha_{y1t}\right] = 0 \\ r_{B1}\cos\alpha_{y1t} - Y_{B1}\left[\dfrac{\pi}{2z} - \dfrac{2x\cos\beta_0\frac{\tan\alpha_0}{\cos\beta_0}}{z} + inv\alpha_{y1t} - inv\left(atan\dfrac{\tan\alpha_0}{\cos\beta_0}\right)\right] = 0 \end{cases}$$

(62)

The resulting system (62) has three unknowns:

- The normal module $m$, which takes standard values;
- The helix angle of the flank $\beta_0$ on the pitch cylinder of the wheel;
- The normal profile shift coefficient $x$ of the wheel.

A further equation is needed to solve the system. To this end, we accept the hypothesis (justified in Appendix A) that the helix angle $\beta_a$ of the flank on the addendum cylinder does not differ substantially from the helix angle from the pitch cylinder $\beta_0$. This remark is extremely important because the slant of the tooth on the addendum cylinder can be measured relatively easily, since the addendum surface exists physically while the pitch surface is fictive. Supposing that the inclination angle of the tooth on the addendum cylinder $\beta_a$ is found, this value is replaced into the first equation of system (62):

$$\frac{m\cos\beta_a}{2Y_{B1}}z\cos\left(atan\frac{\tan\alpha_0}{\cos\beta_a}\right) - \cos\left[asin\left(\frac{r_{B1}}{Y_{B1}}\cos\alpha_{y1t}\right) + \alpha_{y1t}\right] = 0$$

(63)

and then, the equation is solved with respect to $m$:

$$m = 2Y_{B1}\cos\left[asin\left(\frac{r_{B1}}{Y_{B1}}\cos\alpha_{y1t}\right)\right]\frac{\cos\beta_a}{z}\sqrt{1 + \frac{\tan^2\alpha_0}{\cos^2\beta_a}}$$

(64)

The closest standard value is adopted for the normal module:

$$m_n = m_{nstd}$$

and using this value, it is replaced in Equation (63) and then the equation is solved with respect to the angle $\beta_0$:

$$\beta_0 = acos\sqrt{\frac{z^2 m_{nstd}^2}{\left\{2\cos\left[asin\left(\frac{r_{B1}}{Y_{B1}}\cos\alpha_{y1t}\right)\right]Y_{B1}\right\}^2} - \tan^2\alpha_0}$$

(65)

Now, the helix angle on the pitch cylinder is known and this value is replaced in the second Equation (62), obtaining an equation of unknown $x$, the normal profile shift coefficient. The solution is

$$x = \frac{z}{2\tan\alpha_0}\left[\frac{\pi}{2z} + inv\alpha_{y1t} - inv\left(atan\frac{\tan\alpha_0}{\cos\beta_0}\right) - asin\left(\frac{r_{B1}}{Y_{B1}}\cos\alpha_{y1t}\right)\right]$$

(66)

## 3. Exemplification of the Methodology

The methodology we use with the measuring device is as follows: For the teethed wheel to be measured as 1, the axis of measurement is found; it is defined by two spaces between teeth or by two teeth diametrically opposite. For an odd number of teeth, the axis is defined by a space and a tooth placed diametrically opposite. Then, two pairs of balls 2 are introduced in the spaces placed symmetrically with respect to the measuring axis. One of the pairs of balls is brought into contact with the surface of the fixed prism 3, which is assembled by screws on the base plate 4. Next, the mobile prism 5 is carefully moved until its surface meets the second pair of balls. In order to diminish the risk of locking the mobile prism, it has ball bushings which ensure a smooth and noiseless motion along the guiding rods 7, fixed on the base plate by the support parts 8. When all the balls have made firm contact with the prisms, the assembly is fixed using the screw 9. Then, the distance D is measured and the angles $\psi_{up}$ and $\psi_{down}$ are calculated. The methodology is repeated for another configuration of the balls, symmetrically placed with respect to the measuring axis. Next, Table 1 is completed and the relations presented in the paper are applied.

**Table 1.** Measurements results.

| | $r_B$ [mm] | $\psi_{up}$ [deg] | $\psi_{down}$ [deg] | $D$ [mm] | $Y_B$ [mm] |
|---|---|---|---|---|---|
| entry 1 | 5.760 | $2\pi/z$ | 0 | 91.27 | 41.952 |
| entry 2 | 4.75 | $\pi/z$ | $\pi/z$ | 87.51 | 40.008 |

The method is exemplified by applying it to an actual teethed wheel with the number of teeth $z = 14$. The device presented in [25] is used, but the pins were replaced by balls of radii $r_{B1}$ and $r_{B2}$; for the two sets of balls (Figure 14), the dimension over balls $D_1$ and $D_2$ were measured. The values of the involved parameters for two values of the radii at which the balls' centres are positioned— the radii of balls $r_{B1}$ and $r_{B2}$, distances $D_1$ and $D_2$, positioning angles $\psi_{up}$ and $\psi_{down}$—and the values resulting from the calculation with relations (55) and (56) are presented in Table 1.

(a)

**Figure 14.** *Cont.*

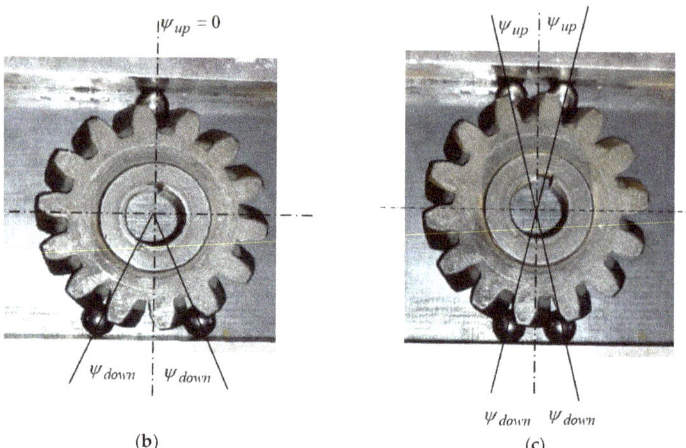



**Figure 14.** (**a**) The experimental device and the two sets of balls. Balls' positions for the measurements: (**b**) $\psi'_{up} = 2\pi/z$, $\psi'_{down} = 0$; (**c**) $\psi''_{up} = \psi''_{down} = \pi/z$.

With these values, a system of transcendental Equation (60) is obtained which is solved using the Newton–Raphson method. The convergence of the solutions of the system as function of the number of iterations is presented in Figure 15. It can be observed that the method is rapidly convergent and conducts to the values of the pressure angles of the involute profile on the circles where the ball–tooth flank contact points are placed.

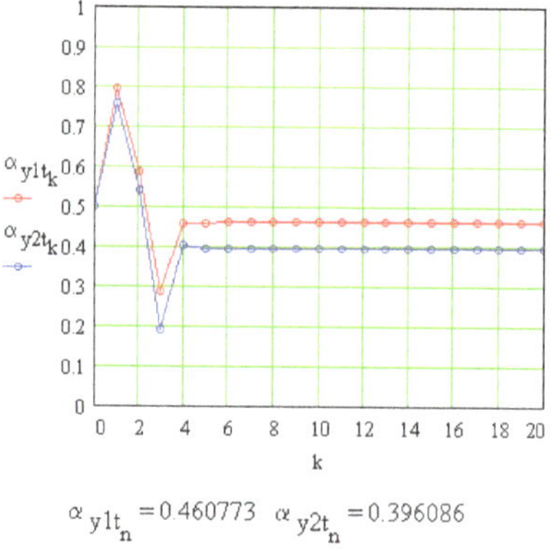



**Figure 15.** The fast convergence of the Newton–Raphson method applied for finding the pressure angles $\alpha_{y1}$ and $\alpha_{y2}$ on the circles of the contact points.

In order to find the helix angle $\beta_a$ of the teeth on the addendum cylinder, two methods were used: first, the wheel was scanned using a laser profilometer (Nanofocus μscan), shown in Figure 16, and then, a photo of the wheel was taken from the side, shown in Figure 17. The helices from the addendum cylinders and the helix angles with respect to the axis of the gear are identified in both images.

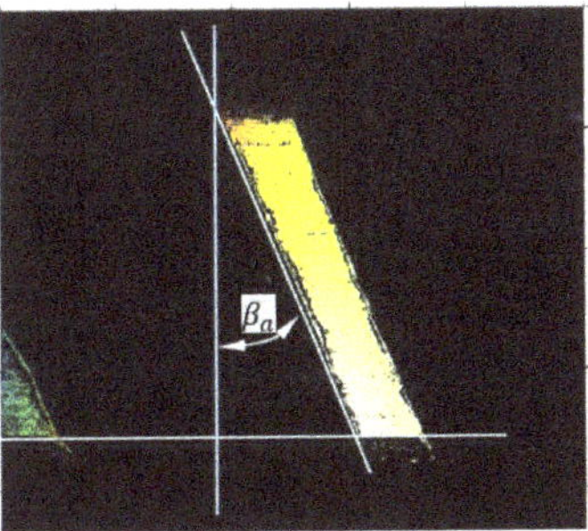



**Figure 16.** The helix angle of the tooth $\beta_a$ on the addendum cylinder, obtained via lateral scan.

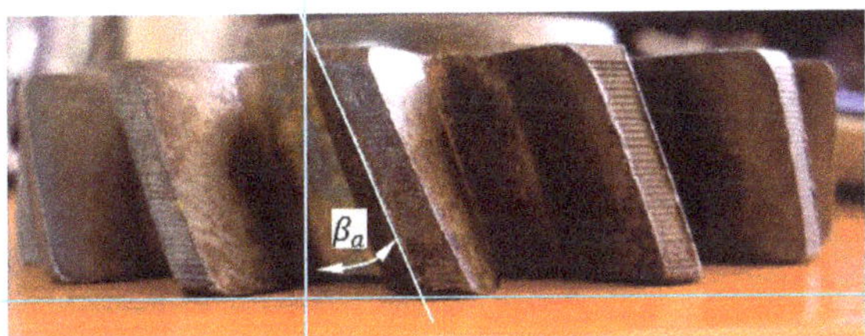



**Figure 17.** The helix angle of the tooth on the addendum cylinder, measured on a lateral photo.

By comparing the measured angles from the two figures, equal values $\beta_a = 23^\circ$ were found. Taking a photo is a much simpler method compared to scanning. Now, with found values $\beta_a$ and $\alpha_{y1t}$ one can apply Equation (64) and the normal module of the gear results in $m = 4.949$. Next, the standard normal module is adopted $m_{std} = 5$; angle $\beta_0$ is obtained by applying relation (65), $\beta_0 = 21.344^\circ$, and next, by applying relation (66), the profile shift coefficient is found $x = 0.105$. Subsequently, it is proven that the values $\alpha_{y1t}$, $\alpha_{y2t}$, $m_{std}$, $\beta_0$, and $x$ verify the system of Equation (59).

## 4. Conclusions

The paper is a progression of a recent work of the authors [25] concerning the accurate determination of the three basic parameters of an actual spur gear with involute teeth (number of teeth, module, profile shift coefficient) where an expedited method and the related measuring device are presented.

The current study introduces a rapid, straightforward, and inexpensive approach for identifying the basic parameters of helical gears. In addition to the three parameters mentioned above, for helical teeth, another basic parameter occurs: the angle of tooth inclination—the helix angle on the pitch cylinder.

For helical gears, due to the inclination of the teeth, the method proposed for the spur gear is inapplicable. Therefore, the rolling bodies introduced between the teeth of the gear to be measured are no longer pins, but balls; thus, the contact with the surfaces of the measuring device is of the point type.

Another consequence of the slanted tooth of the gear is the existence of two profiles: a normal one, where the module has standard values identical to the values of the module of the basic rack, and a transverse one, defined in a section normal to the axis of the wheel. The entire gearing can be studied in the frontal section, similarly to a spur gear, but without the necessity of a standard module. Concerning the control method, the condition that the centre of the ball occupies an imposed position is required. Additionally, we have the condition of positions of the contact points between the balls and the flanks of the teeth. Two sets of balls of different sizes are necessary for a measurement, and the selection of the radii of the balls is made in a manner to ensure the contact point with the flanks on the involute surface of the flank.

Subsequent to the measurements, a system of four equations is obtained, with five unknowns: the pressure angles of the involute profile on the circles passing through the contact point ball–flank, the normal module, the normal profile shift coefficient, and the helix angle on the pitch cylinder.

The research demonstrates that for gears with a typical number of teeth and profile shift coefficient, the helix angle of the tooth on the pitch cylinder can be approximated to be the angle from the top land, which can be conveniently determined through experimental measurements. Therefore, the system becomes compatible and allows for finding, with some approximation, all the basic parameters of the gear. Then, the normal module is adopted as the closest standard value compared to the solution obtained from the system. Further, by employing the standard value for the module, the system of four equations is resolved, leading to the precise determination of the fundamental parameters: the profile's pressure angle at the points of ball–tooth contact, the helix angle on the pitch cylinder, and the normal profile shift coefficient.

The method is then exemplified for a concrete situation.

**Author Contributions:** Conceptualization, SA. and I.D.; methodology, S.A., F.-C.C. and C.B.; software, S.A. and I.-C.R.; validation, I.-C.R. and I.T.; writing—original draft preparation, F.-C.C. and C.B.; writing—review and editing, I.-C.R. and I.T.; visualization, I.T; supervision, I.D., C.B. and F.-C.C. All authors have read and agreed to the published version of the manuscript.

**Funding:** This research received no external funding.

**Data Availability Statement:** Data are contained within the article.

**Conflicts of Interest:** The authors declare no conflicts of interest.

**Appendix A**

The appendix is meant to justify the assumption that for gears with $z > 10$, the helix angle from the addendum cylinder is practically the same as the helix angle from the pitch cylinder.

Next, it is considered a helical gear having the following parameters: number of teeth $z$; normal module $m$; and normal coefficient of profile shift $m$. The helix angle on the addendum cylinder $\beta_0$ is

$$\beta_a = atan\left(\frac{d_a}{d}tan\beta_0\right) \tag{A1}$$

where $d$ is the pitch diameter:

$$d = z\frac{m}{cos\beta_0} \tag{A2}$$

and $d_a$ is the addendum diameter:

$$d_a = d + 2(h_{a0}^* + x)m = m\left(\frac{z}{cos\beta_0} + 2h_{a0}^* + 2x\right) \tag{A3}$$

where $h_{a0}^* = 1$. Then, the following equation is obtained:

$$\beta_a = atan\left(\frac{m\left(\frac{z}{cos\beta_0} + 2h_{a0}^* + 2x\right)}{z\frac{m}{cos\beta_0}}tan\beta_0\right) = atan\left(\frac{\frac{z}{cos\beta_0} + 2h_{a0}^* + 2x}{z}sin\beta_0\right) \tag{A4}$$

The number of teeth and the profile shift coefficient affect the helix angle on the addendum cylinder. These aspects are presented in Figures A1 and A2 for $z = 10 \div 80$.

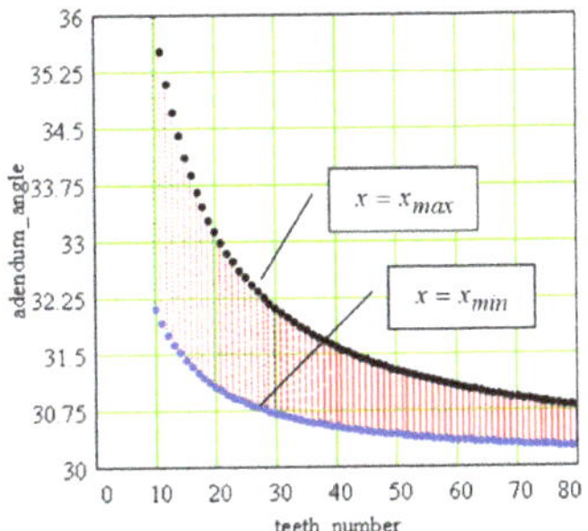


**Figure A1.** The helix angle on the addendum cylinder versus the number of teeth.

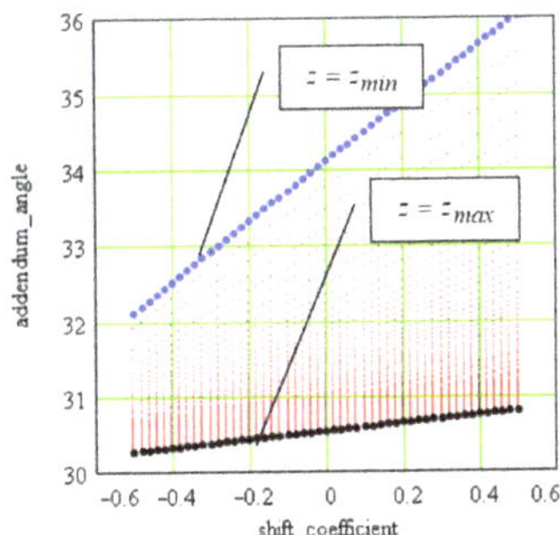


**Figure A2.** The helix angle on the addendum cylinder versus the profile shift coefficient.

## References


1. Freudenstein, F. An Analytical Approach to the Design of Four-Link Mechanisms. *ASME Trans.* **1954**, *76*, 483–489. [CrossRef]
2. Flores, P.; Ambrósio, J.; Pimenta Claro, J.C.; Lankarani, H.M. *Kinematics and Dynamics of Multibody Systems with Imperfect Joints*; Springer: Berlin/Heidelberg, Germany, 2008; pp. 23–45.
3. Seherr-Thoss, H.C.; Schmelz, F.; Aucktor, E. *Universal Joints and Driveshafts: Analysis, Design, Applications*, 2nd ed.; Springer: Berlin/Heidelberg, Germany, 2006; pp. 53–79, 109–245.
4. Dudiţă, F. *Mobile Homokinetic Couplings*; Tehnică: Bucharest, Romania, 1974; pp. 209–225. (In Romanian)
5. Innocenti, C. The instantaneous transmission ratio of a driveshaft composed of a tripod joint and a fixed constant velocity joint. *Mech. Mach. Theory* **2023**, *189*, 105430. [CrossRef]
6. Qiu, Y.; Shangguan, W.B.; Lou, Y. Kinematic analysis of the double roller tripod joint. *J. Multi-Body Dyn.* **2020**, *234*, 147–160. [CrossRef]
7. Pennestrì, E.; Rossi, V.; Salvini, P.; Valentini, P.P.; Pulvirenti, F. Review and kinematics of Rzeppa-type homokinetic joints with straight crossed tracks. *Mech. Mach. Theory* **2015**, *90*, 142–161. [CrossRef]
8. Alaci, S.; Doroftei, I.; Ciornei, F.-C.; Romanu, I.-C.; Doroftei, I.-A.; Ciornei, M.-C. A new RP1PR type coupling for shafts with crossed axes. *Mathematics* **2023**, *11*, 2025. [CrossRef]
9. Radzevich, S.P. *Theory of Gearing. Kinematics, Geometry and Synthesis*, 3rd ed.; CRC Press: Boca Raton, FL, USA, 2023; pp. 79–107.
10. Phillips, J. *General Spatial Involute Gearing*; Springer: Berlin/Heidelberg, Germany, 2003; pp. 41–61.
11. Reuleaux, F. *The Kinematics of Machinery: Outlines of a Theory of Machines*; Dover Publications: Mineola, NY, USA, 2012; pp. 121–157.
12. Alaci, S.; Ciornei, F.-C.; Doroftei, I. The conjugate profile of the circular teeth of a spur gear. Part II: Problem solution. *IOP MSE* **2020**, *997*, 012068. [CrossRef]
13. Boral, P.; Gołębski, R.; Kralikova, R. Technological Aspects of Manufacturing and Control of Gears—Review. *Materials* **2023**, *16*, 7453. [CrossRef] [PubMed]
14. Ling, M.; Ling, S.; Li, X.; Shi, Z.; Wang, L. Effect on the measurement for gear involute profile caused by the error of probe position. *Meas. Sci. Technol.* **2022**, *33*, 115013. [CrossRef]
15. Palermo, A.; Britte, L.; Janssens, K.; Mundo, D.; Desmet, W. The Measurement of Gear Transmission Error as an NVH Indicator: Theoretical Discussion and Industrial Application via Low-Cost Digital Encoders to an All-Electric Vehicle Gearbox. *Mech. Syst. Signal Process.* **2018**, *110*, 368–389. [CrossRef]
16. Yuan, B.; Chang, S.; Liu, G.; Wu, L.Y. Quasi-Static and Dynamic Behaviors of Helical Gear System with Manufacturing Errors. *Chin. J. Mech. Eng.* **2018**, *31*, 30. [CrossRef]

17. Colbourne, J.R. *The Geometry of Involute Gears—Softcover Reprint of the Original*, 1st ed.; Springer: New York, NY, USA, 1987; pp. 191–200, ISBN 978-1-4612-4764-7.
18. Jalaska, D. *Gears and Gear Drives*; John Wiley & Sons Ltd.: West Sussex, UK, 2012; pp. 106–155. ISBN 9781119941309.
19. Jantzen, S.; Neugebauer, M.; Meeß, R.; Wolpert, C.; Dietzel, A.; Stein, M.; Kniel, K. Novel measurement standard for internal involute microgears with modules down to 0.1 mm. *Meas. Sci. Technol.* **2018**, *29*, 125012. [CrossRef]
20. Härtig, F.; Stein, M. 3D involute gear evaluation—Part I: Workpiece coordinates. *Measurement* **2019**, *134*, 569–573. [CrossRef]
21. Stein, M.; Härtig, F. 3D involute gear evaluation—Part II: Deviations—Basic algorithms for modern software validation. *Meas. Sci. Technol.* **2022**, *33*, 125003. [CrossRef]
22. Guo, X.; Shi, Z.; Yu, B.; Zhao, B.; Li, K.; Sun, Y. 3D measurement of gears based on a line structured light sensor. *Precis. Eng.* **2022**, *61*, 160–169. [CrossRef]
23. Stein, M.; Keller, F.; Przyklenk, A. A unified theory for 3d gear and thread metrology. *Appl. Sci.* **2021**, *11*, 7611. [CrossRef]
24. Bolotovskaia, T.F.; Bolotovski, I.A.; Smirnov, V.E. *Guide for Gear Profile Shifting*; Institutul de Documentare Tehnică: Bucharest, Romania, 1964. (In Romanian)
25. Alaci, S.; Ciornei, F.-C.; Romanu, I.-C.; Doroftei, I.; Bujoreanu, C.; Tamaşag, I. A New Direct and Inexpensive Method and the Associated Device for the Inspection of Spur Gears. *Machines* **2023**, *11*, 1046. [CrossRef]

MDPI

St. Alban-Anlage 66

4052 Basel

Switzerland

www.mdpi.com

*Applied Sciences* Editorial Office

E-mail: applsci@mdpi.com

www.mdpi.com/journal/applsci